GW01337285

An Introduction to Stirling Engines

JAMES R. SENFT

Moriya Press

Copyright © 1993 James R. Senft

All rights reserved. No part of this book may be reproduced without written permission.

ISBN 0-9652455-0-0

Fifth Printing 2000

Cover illustration by Josip Antonic´

An Introduction to Stirling Engines

Great are the works of the Lord
To be studied by all who delight in them

Psalm 111

The Stirling engine is a fascinating device. One end is heated, the rest is kept cool, and useful work comes out through a rotating shaft. It is a closed machine with no intake or exhaust. Heat is applied from the outside. Anything that will burn will serve to run a Stirling engine: coal, wood, straw, gasoline, kerosene, alcohol, propane, natural gas, methane and so on. But combustion is not required. Only heat is needed to bring a Stirling engine to life. Stirlings will operate on solar energy, or geothermal energy, or on surplus heat from industrial processes.

Invented in 1816, the first Stirling engines were large machines working in industrial settings. Later, smaller versions of these safe and quiet engines became popular for domestic and light industrial chores such as driving fans, sewing machines, and water pumps. The early Stirling engines contained ordinary air as the working substance and so were popularly known as "hot air" engines. The captive air was cyclically subjected to heating, expansion, cooling, and compression by the movement of the various parts of the engine.

Small hot air engines remained in production and use into the early 1900's, but were superseded as the internal combustion engine improved and electrification advanced. At the present time, no practical size Stirling engines are in production, but they are the subject of considerable research and development efforts because of their potential for high efficiency and clean and quiet operation. Current experimental versions of the Stirling engine have achieved impressive levels of performance. Using high temperature alloys, new mechanical drives, computer aided heat exchanger designs, and charged with helium or hydrogen at high pressure, modern Stirlings can easily surpass typical gasoline and small diesel engines in efficiency and in power-to-weight ratio. In quietness and low pollution levels, Stirling engines have no serious competitors. In the future this can mean cleaner automobiles, quiet lawnmowers, and electricity generated from the sun.

Heat Engines

The Stirling is a heat engine. To fully understand how it works and its potential uses, it is essential to know its place in the general domain of heat engines. A heat engine is a device which can continuously convert thermal energy, or heat, into mechanical energy, or work. Heat is supplied or applied to the engine in one way or another and useful work is produced. As long as heat is supplied, an engine produces mechanical energy.

A classic example of a heat engine is the steam locomotive of yesteryear. Thermal energy was supplied by burning coal or wood. This heated a water-filled boiler and generated high pressure steam. The steam did work by expanding in cylinders and pushing on pistons to drive the wheels and pull the train along. After each piston stroke, spent steam, still a bit warm, was exhausted in a puff up the chimney to make room for a fresh charge of high pressure steam into the cylinder. It was heat that made the old locomotives chug along.

And heat still does, but by means of a different engine, the diesel. The diesel is a form of internal combustion engine, the same general type of engine that powers automobiles and lawnmowers. In the internal combustion engine, heat is supplied by burning fuel, usually a liquid, such as gasoline or light oil, inside part of the engine. Some of the heat is converted to work by expanding the hot gaseous combustion products against a piston. The rest of the heat is rejected in the exhaust, radiator, or cooling fins of the engine. As long as there is fuel to supply heat, the engine does useful work.

These three processes of heat absorption, work production, and heat rejection are common to all heat engines. As in the examples already mentioned, they are an integral part of the operation of every heat engine. The laws of nature require these processes, and why this is so can be most clearly seen by looking at a very simple form of heat engine.

An Elementary Heat Engine
Consider a closed-ended cylinder fitted with a piston as shown in Fig. 1 with some ordinary air captured inside. The piston is freely movable but practically airtight. Suppose that initially the whole device is at a cool room temperature. This includes the air inside which also is assumed to be at the same pressure as the normal atmospheric air outside. Given these conditions, the piston will just stay where it is.

Work can be obtained from this device by heating the cylinder and raising its contents to a higher temperature. A flame can be applied, or the sun's rays can be focused onto the cylinder, or the end can simply be dunked into a bucket of hot water. The temperature of the air inside will increase and so will its pressure. The higher pressure can then push the piston outward and do useful work against a resistance. Any heat source will give some work, but the higher the heat source temperature, the greater will be the work that can be obtained.

Converting heat into work like this on a one-time only basis is easy, but a heat engine must be capable of continually producing work. Work can be gotten out of a heated air-filled cylinder only as long as expansion of the air inside can occur. If the piston is allowed to move out too far, it will pop out of the cylinder and that will be an end to getting any more work! The whole thing would have to be cooled down and put back together before it could do useful work again. So the expansion process must be stopped before this happens. If the cylinder was made really long, the expansion could go out farther, but still there would be a limit. It would only go out until the air pressure inside drops all the way back down to atmospheric pressure. That would be it for work production.

So either way, work can be extracted from a single source of heat by a single device only in a limited amount. The limit occurs when

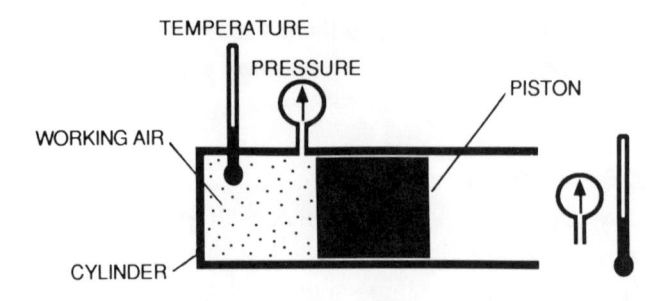

Fig. 1 A cylinder containing air (or other gas) held captive by a movable piston can be used to convert heat into work. Initially the air inside is at the same temperature and pressure as the air outside.

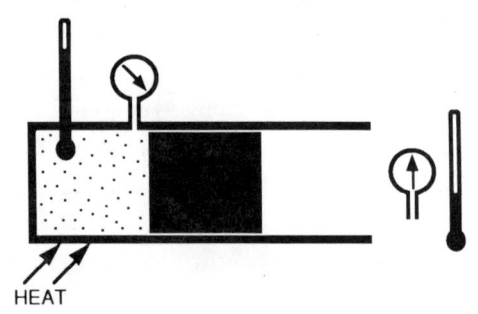

Fig. 2 The cylinder can be heated to bring it up to a higher temperature. The pressure of the air inside rises above atmospheric in proportion to the temperature increase.

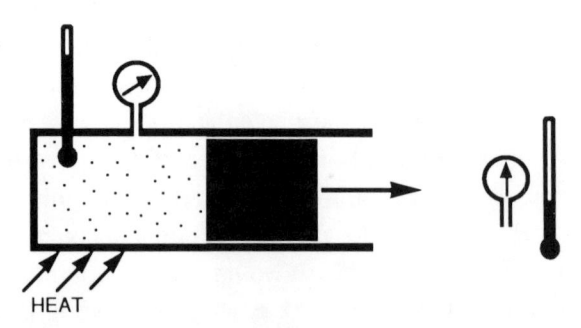

Fig. 3 The higher pressure air inside can then push the piston outward, doing work. Heat can be continually applied during this expansion to keep the temperature up at the peak to get as much work out as possible. Pressure steadily decreases in inverse relation to the increasing volume.

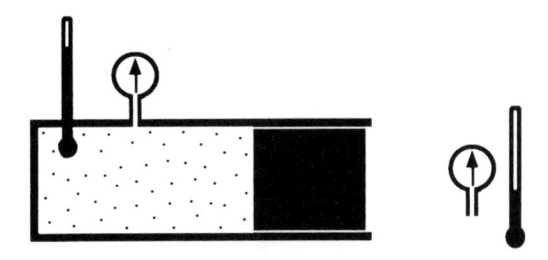

Fig. 4 Expansion can be carried out until the pressure inside has decreased all the way back to atmospheric. If heat continued to be added through all of the expansion process, the temperature will still be at the peak.

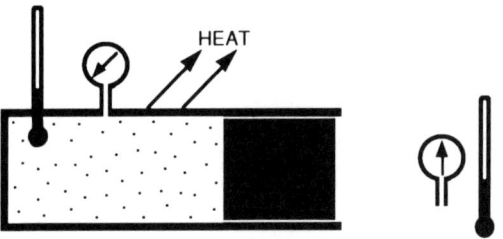

Fig. 5 With the heat source removed, the cylinder cools back down to the surrounding room temperature. The pressure of the air inside drops below atmospheric because of the expanded volume.

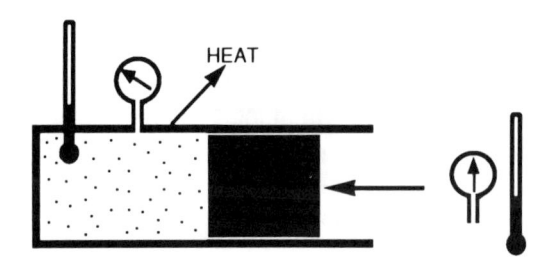

Fig. 6 The piston will be pushed inward by the higher outside pressure. Continued cooling maintains the low temperature of the cylinder. Pressure rises as compression stroke proceeds. The compression stroke ceases when the internal pressure has risen back up to atmospheric. If no leakage past the piston has occurred, the piston and the air inside will end up at exactly the same state as initially in Fig. 1, completing a thermodynamic cycle.

the device or the working substance reaches a state beyond which it cannot go. In this example the limit is determined by the temperature of the heat source and the length of the cylinder, or by the atmospheric pressure outside the cylinder.

To get the most work out of the cylinder unit under consideration, suppose that as expansion occurs heat is continually applied so that the temperature inside the cylinder is as hot as the source that is being used can make it. Ideally this would be the temperature of the source itself. This will keep the pressure inside as high as possible. Of course, as the expansion proceeds the pressure inside will decrease, but as long as the inside pressure is higher than the outside, the piston will be pushed outward to do work. So to get all that can be gotten, the cylinder could be made long enough so that the expansion can continue until finally the inside pressure just equals the outside. This will give the most work out using only the given cylinder and the one heat source. The steps so far are shown in Figs. 1 - 4.

In order to get any more work from this device, it is necessary to somehow return it to its original condition so that the process can be repeated. In returning it to its initial state, the process becomes what is called a cyclic process or simply a cycle. Cycles can be performed over and over. Therefore a cyclic device can produce work indefinitely. This is precisely why heat engines utilize cycles. By carrying a working substance such as air through a repeating cycle, heat engines are able to continually convert heat into work. For this reason, cycles are the heartbeat of heat engines.

To restore the cylinder unit from the condition in which it was left after the expansion in Fig. 4, begin by removing the heat source and letting the unit cool back down to room temperature, Fig. 5. The pressure will drop, going below atmospheric because the volume is larger than it was initially and it is again at room temperature. The difference in pressure across the piston will move it inward. In fact the device can now deliver more work to an external resistance. Of course it will do so in the reverse direction during the decrease in volume or compression, but it is still useful work. In most engines a mechanism linking the piston to the output shaft automatically reverses so that the engine continuously delivers work in one direction. During the compression, Fig. 6, the cylinder is allowed to lose or give up more thermal energy to the room air. The more effectively the cylinder can lose heat, the more work will be gotten out during the compression process. In fact the best that can be done is to keep the air inside just about at room temperature throughout the compression stroke. The idea here is analogous to that described for the expansion process, and will likewise give the most work.

As compression proceeds, pressure within the cylinder rises, even though the temperature is held down to the minimum. This process ceases when the pressure inside has risen back up to atmospheric. If the piston fit is perfect with no leakage of air in or out, the piston will return to its exact original position and the air inside will be at its initial state, just as in Fig. 1. Thus the air inside has been taken through a complete thermodynamic cycle, and the device can function as the basic component of a heat engine.

Usually a heat engine is considered complete with a mechanism which converts the reciprocation of the piston into uniform rotary motion. Rotation is a convenient form of output for driving common implements of work such as pumps, saws, fans, and conveyors. Figure 7 shows the common crank and connecting rod mechanism fitted to the piston of the engine cylinder discussed above. The flywheel shown there is essential for a single cylinder engine. Its momentum keeps the machine moving along when the crank passes by the points of reversal or dead center positions.

Another general feature of an engine mechanism is automatic working. The mechanism functions to move and regulate everything so that the engine will automatically repeat its thermodynamic cycles. This requires somehow making the engine working fluid absorb or reject heat at the right times. In the basic engine cylinder unit described in Figs. 1-6 above, this was manually done by applying or removing the heat source. A way to do this automatically which Rube Goldberg might conjure up is shown in Fig. 8. The engine there,

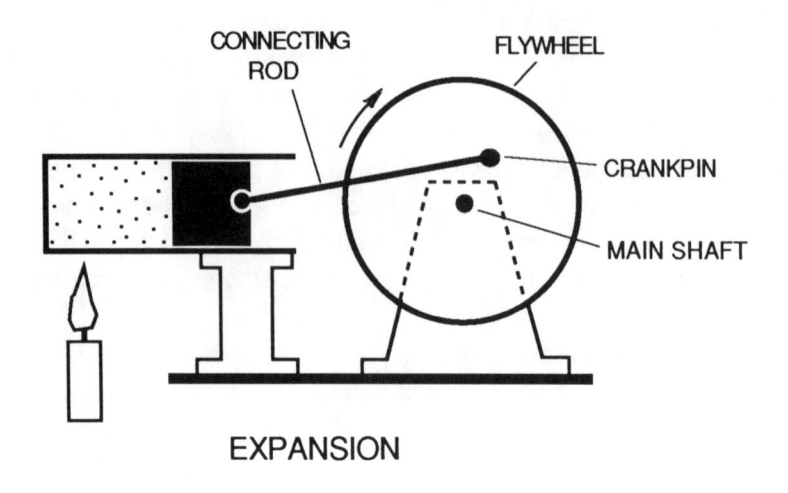

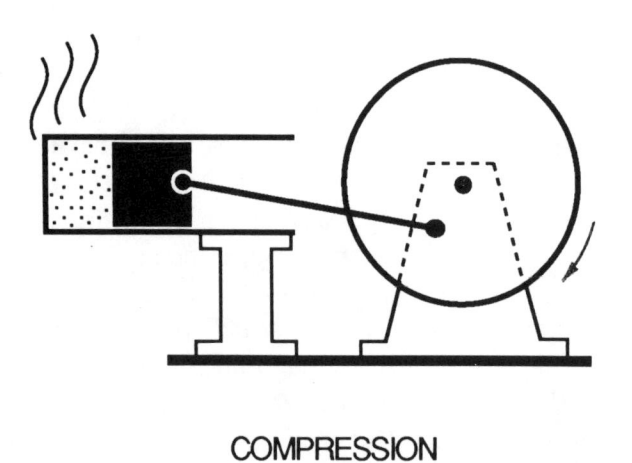

Fig. 7 A crank and connecting rod mechanism converts reciprocating piston motion into continuous rotary shaft motion.

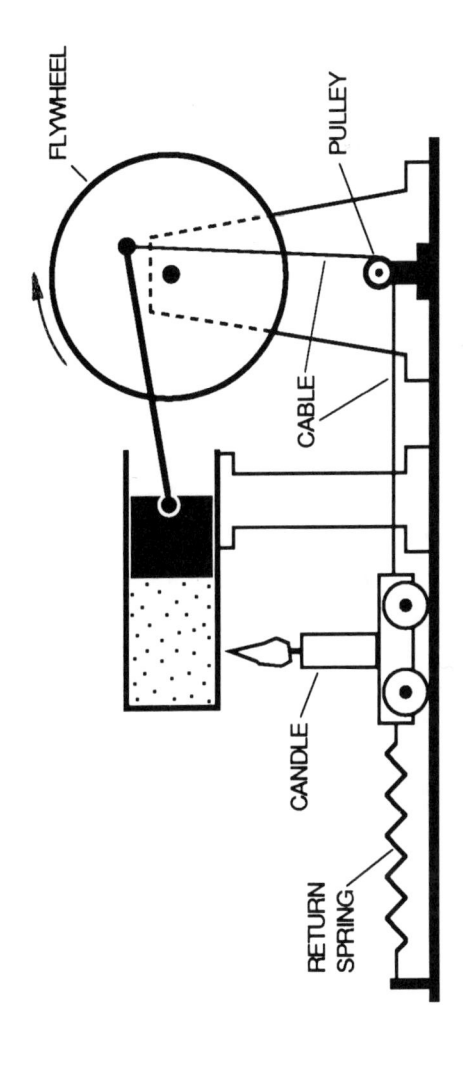

Fig. 8 "Moover", a whimsical approach to making a complete automatically operating heat engine.

which will be referred to as "Moover", has a candle flame as the heat source. A cable system is set up to pull the candle under the cylinder just when the engine is performing its expansion, and then to let the spring retract it out of the way for cooling. Moover would indeed work in principle, but would be exceedingly slow and inefficient. It is what one might think of first, a workable but whimsical contraption. Although it is not the way to make a practical engine, it does illustrate all of the features of automatically operating heat engines in a simple and clear enough way to see their essential characteristics.

Slightly better is the multicylinder version of Moover shown in Fig. 9. A number of identical cylinders are fastened together in a radial assembly which rotates about a stationary shaft. The shaft carries a fixed crank linked to the pistons by connecting rods. As the cylinder assembly rotates, the pistons move in their cylinders in the normal reciprocating fashion. In this version, the cylinders dip into a trough of hot water at the bottom of their circle of rotation, and then cool during the rest of the revolution in the surrounding air. The engine could also be heated by combustion or arranged with the upper half of it shaded so that it will operate from solar energy. The crank is positioned so that expansion takes place when each cylinder is hot, and compression when cool. Each cylinder goes through the same cycle of heating, expansion, cooling, and compression exactly as in Moover. However, there is always at least one hot cylinder somewhere in its expansion stroke so that rotation would be continuous and no flywheel would be needed.

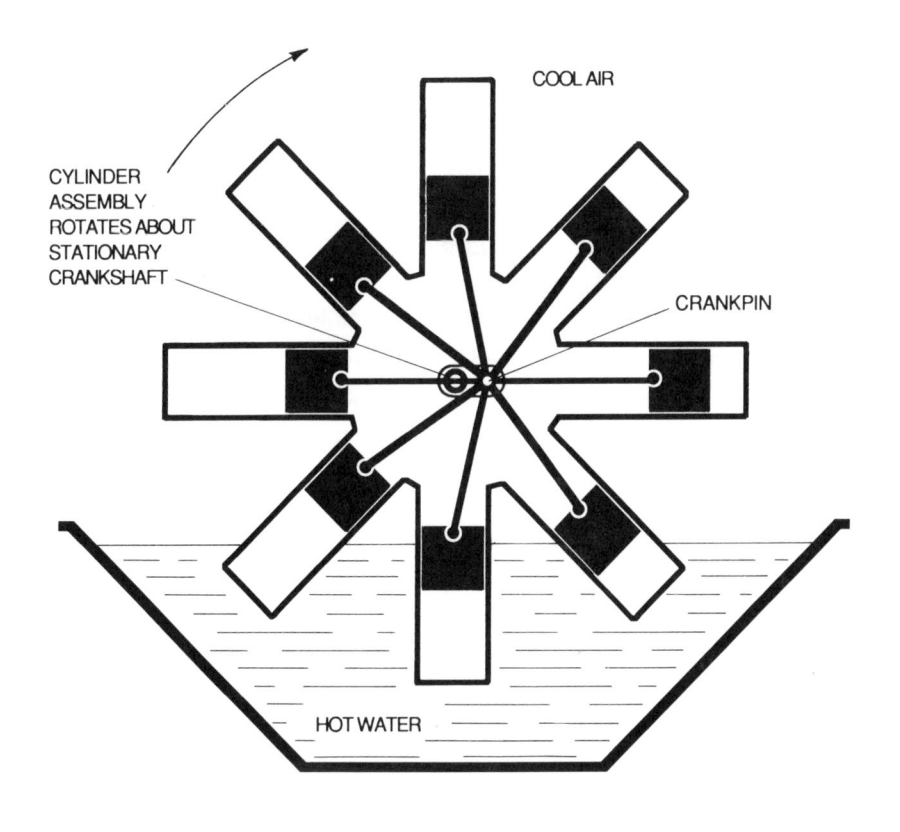

Fig. 9 A multicylinder engine based on the same cycle as Moover.

General Characteristics of Heat Engines and the Laws of Thermodynamics

General insights about all heat engines can be made based upon the elementary but essentially complete engine described above. First, in order to make Moover function as a heat engine, that is, to make it carry out a thermodynamic cycle, it was necessary to have two temperature extremes between which to work. A source or reservoir of relatively high temperature thermal energy was needed to initiate the cycle. The working air absorbed heat from this high temperature reservoir during the expansion process of the cycle. Then, to effect the cooling, the engine had to give up heat. The surrounding room air absorbed it and thus it served as the heat sink or lower temperature reservoir. This is a requirement of all possible engine cycles. Every engine cycle requires heat reservoirs at two different temperatures. This is a simple statement of a principle of physics called the Second Law of Thermodynamics. Every engine needs a hotter reservoir for the heat source and a cooler one for the heat sink. Combustion of fuel produces hot gases which form the heat source of most current practical engines. In fact some will only work by burning a specific fuel, but others, like our simple Moover theoretically can run from any heat source. The heat sink that is also needed by every engine is invariably atmospheric air or water drawn from a river or lake.

An abstract or purely conceptual representation of the general heat engine is shown in Fig. 10. The rectangles represent the source and sink reservoirs at high and low temperatures of T_H and T_C respectively. The circle represents the engine itself. Q_i is the heat supplied to the engine each cycle and Q_o is the heat rejected by the engine each cycle. W stands for the work output per cycle.

A second common characteristic of heat engines is a dimensional change in the working substance. It is through a dimensional change that work is taken out of or put into the working substance. Most practical engines use a gas or mixture of gases like air for example for the working substance in which case the dimensional change is one of volume, an expansion and contraction.

This change in volume is effected by a movable piston, diaphragm, bellows, or equivalent device. A change in volume corresponds to movement or displacement of the piston, and change of position is one of the two principal components of work. The other is force. Force acting over a distance is work. The force in an engine arises from the gas pressure acting on the piston area.

In a part of every engine cycle, an expansion or volume increase takes place, and in another part a volume reduction or compression occurs. In the expansion, the engine gas delivers work to the piston and usually this goes into a flywheel by means of mechanical linkage, like the crank and connecting rod in Moover. Some of this is taken to do useful work outside of the engine, but in some engines, not all of it can be used up. In a diesel engine for example, where the compression ratio is on the order of 15 to 1, a considerable amount of flywheel energy is needed to carry out the compression process. However, in every working engine, more work is produced in the expansion than is needed for the compression. The difference is the potentially usable or net work of the engine cycle.

The origin of this surplus energy is thermal. The energy content of the heat absorbed from the hotter reservoir is greater than that rejected to the lower temperature reservoir and the surplus appears as work. This is the so called First Law of Thermodynamics which for engines can be stated in the following way: The net work that is done by an engine over a complete cycle is the difference between the heat absorbed from the hot source and the heat rejected to the cold sink. In the symbolism of Fig. 10, $W = Q_i - Q_o$. In summary then, all heat engines must operate in a cycle, and they must function by absorbing heat from a source, rejecting some at a lower temperature, and converting the difference into work.

Efficiency and the Second Law

How good an engine is depends upon whether it will produce enough power to do its work quickly, how big and heavy it is, how reliable, how hard it is to make, and how efficient it is at converting

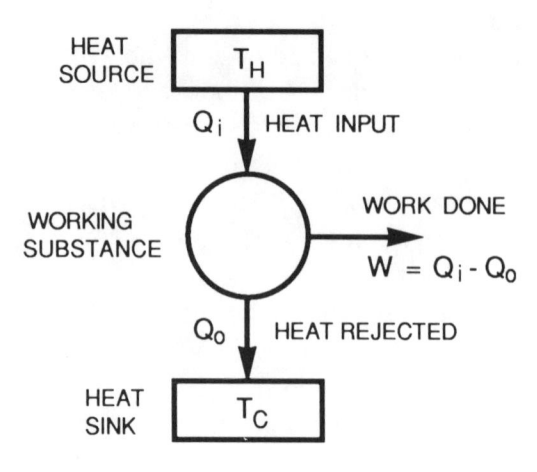

Fig. 10 A general heat engine diagram showing thermal energy absorbed from the high temperature reservoir, heat rejected to the lower temperature reservoir, and the difference converted into work.

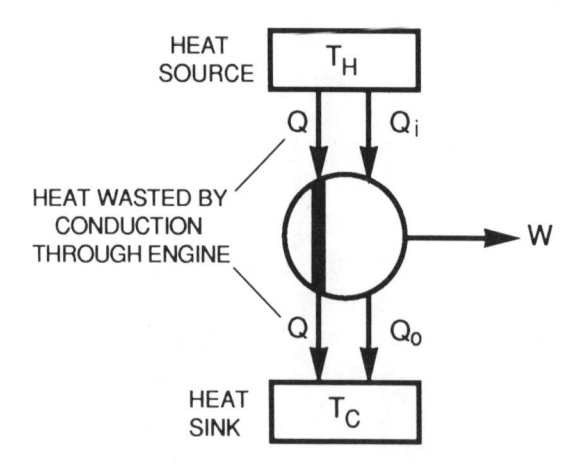

Fig. 11 A conduction path through an engine reduces efficiency because more heat is taken from the source but the same amount of work is produced by the engine.

heat to work. Some of these considerations are highly dependent upon the use of the engine. What is important in one application may not be in another. But in nearly all, efficiency is pretty important. No one wants to waste fuel or burn more than is reasonable. Efficiency is the measure of how much work is obtained for each bit of heat or fuel used. It is the ratio of the work an engine produces to the heat input needed to produce it; in symbols: W/Q_i. Efficiency is often expressed as a percentage. Typical automobile engines are about 25% efficient, small diesel engines up to 35%. The latest very large diesel ship engines can be made up to 50% efficient, and these are the most efficient prime movers presently available.

Now one would expect less than 100% efficiency in a real engine. After all, there are always imperfections in real engines. Even the best pistons do not seal perfectly, valves leak, belts slip, and friction is present everywhere parts move against each other. But on top of all these shortcomings, the laws of thermodynamics impose an even more fundamental limitation. As was seen above, the Second Law always requires some heat be rejected to a lower temperature reservoir. Thus all of the heat absorbed from the higher temperature source cannot be converted to work. Since the difference is the potential work that can be gotten out of the engine, this means that efficiency can never be 100% for any engine, no matter how well designed and built, even if it were perfect.

It may at first seem strange that no engine, not even an ideal engine, can convert all the heat it absorbs into work. But this is what the two laws of thermodynamics imply. The absolute best efficiency that can be obtained depends upon the two temperatures between which the engine operates. It is known as the Carnot efficiency after the early eighteenth century engineer who first proved its existence. The exact formula for it is

$$\text{CARNOT THERMAL EFFICIENCY} = \frac{T_H - T_C}{T_H}$$

where the temperatures are expressed in an absolute scale, such as Kelvin. In other words, the efficiency of any engine cannot exceed the difference in the source and sink temperatures divided by the source temperature. So for example, if an engine is using a source at 900 K (steel is red hot at this temperature) and the sink temperature is 300 K (a little above room temperature), then the best efficiency the engine can give is

$$\frac{900-300}{900} = \frac{600}{900} = \frac{2}{3}, \text{ or } 67\%.$$

The formula shows that the farther apart the temperatures are, the higher is the potential efficiency of the engine. So if you have an engine running and can manage to crank up the heat and get it hotter, it can run more efficiently. It will usually produce more power also. Loosely speaking, everything gets better when the temperatures of the heat source and sink are farther apart. But it must be kept in mind that the Carnot efficiency is the theoretical upper limit on performance. Even it cannot be attained because of all the unavoidable defects in a real engine. In fact a lot of care has to be exerted when making and operating an engine to get a decent fraction of the Carnot fraction!

The first step in obtaining good efficiency is avoidance of direct "heat wasters" or thermal "short circuits". These are features in the engine or the cycle that cause heat to be passed from the hot to the cold reservoir without contributing to the work done by the engine gas. The simple engine Moover described above has a really good cycle as far as the air inside is concerned. But the cylinder material has to be heated and cooled each cycle to make air inside heat up and cool down also. Each cycle, part of the heat absorbed by the engine prior to the expansion stroke goes to warming up the cylinder material. At the end of expansion, this same heat is rejected to the sink to cool the cylinder material down. This thermal energy does not produce any work, it just goes into the engine metal and then back out again as Fig. 11 indicates schematically. Only heat exchanged with the air inside the engine can produce work, the rest is just wasted.

This kind of direct heat waste can drastically reduce engine efficiency. If the cylinder is made fairly substantially to withstand the rigors of usage, it will have a high heat capacity relative to the air inside, so a lot of heat can be short circuited through the engine. The ideal solution to this problem is one where the air inside could be heated and cooled without cyclically heating and then cooling the cylinder material. Fortunately, this can actually be done by means of an ingenious device called a displacer.

The Displacer

Imagine a completely closed cylinder as shown in Fig. 12 containing some air and a cylindrical "plunger" about half the length of the cylinder. The plunger is a fairly loose fit in the cylinder so that tilting or shaking the cylinder can move it from one end to the other. The plunger occupies space inside. Wherever it happens to be, the air isn't. So if it is at the left hand end, the air inside is for the most part in the right hand end, and vice-versa. Moving the plunger moves or displaces the air in the cylinder. For this reason, the plunger is usually referred to as a displacer.

Now suppose one end of this cylinder is heated and the other end is kept cool. When the displacer is all the way into the hot end as in Fig. 13, the air is in the cold section, so it is cool and relatively low in pressure. When the displacer is moved into the cold end, the air flows around it and ends up in the hot section as shown in Fig. 14. The air temperature rises and its pressure increases. Thus moving the displacer back and forth causes the air to be alternately heated and cooled. The beauty of the displacer is that the cylinder can be maintained at steady temperatures - one end always hot and the other always colder. This not only reduces wasted heating and cooling of the same section of the cylinder over and over, but it also permits much more rapid heating and cooling of the air inside.

The usual means of moving a displacer is via a slender rod attached to the displacer and passing through a close fitting bushing in the cold end of the cylinder. This idea is shown in Fig. 15. This rod

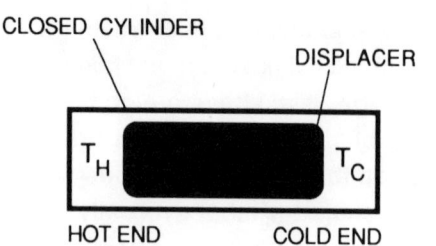

Fig. 12 An air-filled cylinder containing a displacer. One end of the cylinder is heated and the other end is cooled.

CLOSED CYLINDER   DISPLACER

T_H T_C

HOT END   COLD END

Fig. 13 When the displacer is in the hot end, most of the air inside the cylinder is in the cold section. The air is cool and its pressure is relatively low.

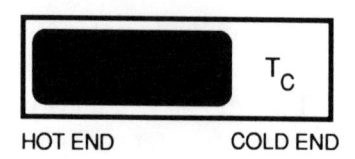

T_C

HOT END   COLD END

Fig. 14 When the displacer is in the cold end, the air is in the hot end. The air temperature and pressure are higher.

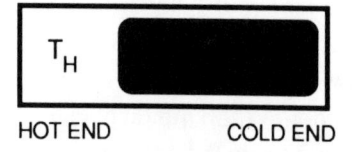

T_H

HOT END   COLD END

Fig. 15 To move the displacer back and forth, it can be attached to a slender rod which passes through a close-fitting practically airtight bushing in the cold end of the cylinder.

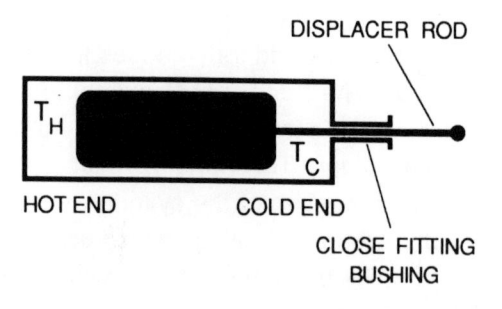

DISPLACER ROD

T_H
T_C

HOT END   COLD END

CLOSE FITTING BUSHING

18

also supports the displacer and keeps it away from the cylinder walls so it can freely move without scraping. The annular space around the displacer also permits a uniform flow of air past it as it reciprocates.

The Stirling Thermodynamic Cycle

This displacer-cylinder unit can be used in conjunction with a piston-cylinder unit to yield an engine with a good cycle and much better efficiency than our rudimentary engine Moover described above. The best combination of the two units from a thermodynamic point of view is one in which the piston cylinder is just an extension of the displacer cylinder. But a simpler configuration is that shown in Fig. 16. The piston-cylinder unit is connected by a short passage to the displacer-cylinder unit.

With this arrangement, known as the "split-cylinder" or "gamma" configuration, it is possible to carry the air inside through a complete thermodynamic cycle with steady heating of one end and steady cooling of the other. The cycle is thermodynamically identical to the one described in connection with Moover, but now can be done more efficiently and rapidly. The four steps are carried out as shown in Figs. 16 a- d.

In Fig. 16a, the piston is in as far as it can go and the displacer is moving from the hot end into the cold end. Essentially all of the engine air ends up in the hot section of the displacer cylinder. Of course there is some air in the connecting passage and some more in the annular region around the displacer, but most of the air will be in the hot end. Its temperature and pressure rise and all is ready for a productive expansion stroke of the piston.

The energy needed to move the displacer is small. The friction is only that of the rod sliding in its bushing and the internal air flowing around the displacer. Although small, the energy to overcome this friction work must come from somewhere. It is usually supplied from a rotating flywheel by means of linkage not shown here. Various mechanical displacer and piston drive linkages will be described later. In this first step, the engine air was heated while the total engine

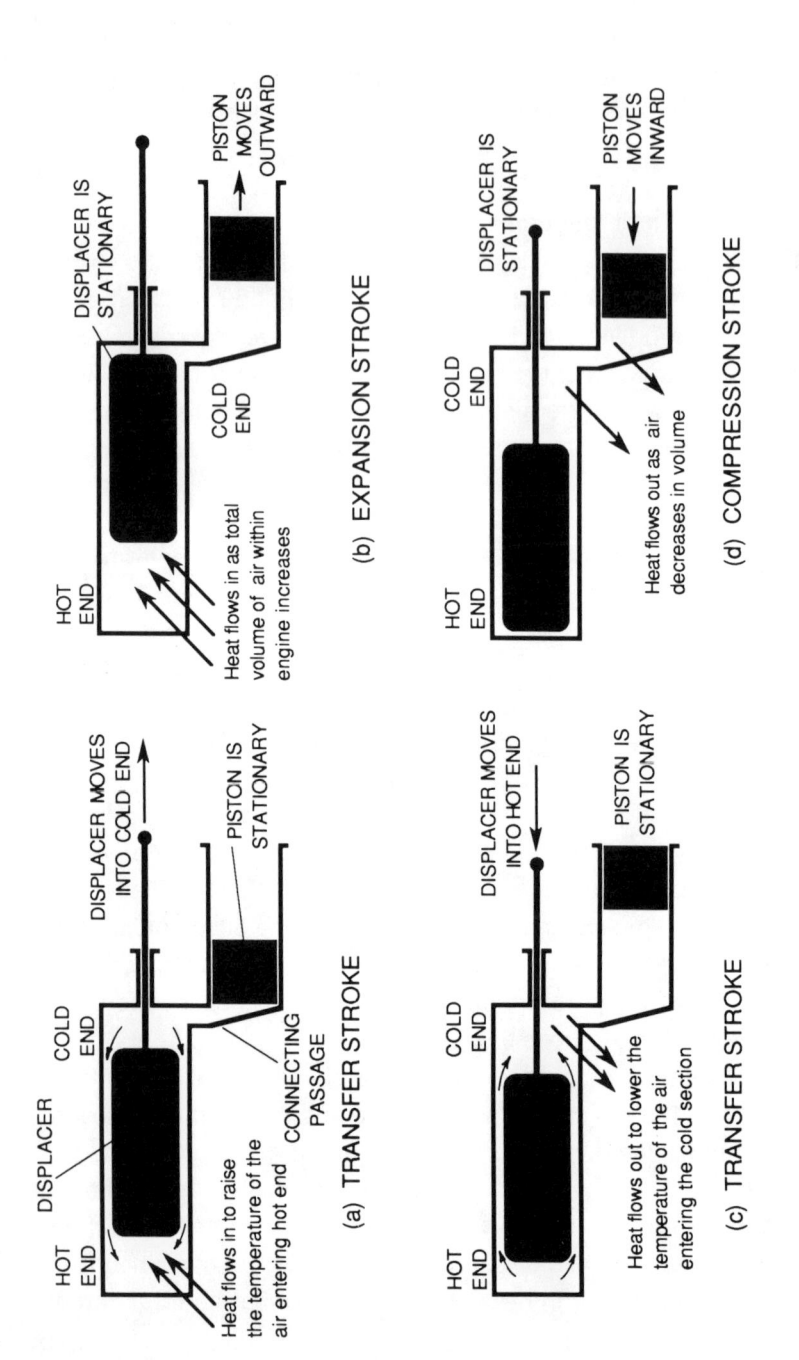

Fig. 16 A displacer cylinder connected to a piston cylinder can rapidly carry out a simple version of the four-step Stirling thermodynamic cycle.

volume was fixed. This first step of the cycle is a constant volume heating of the engine air. The result is hotter higher pressure air inside the engine.

In the next step of the cycle shown in Fig. 16b, the displacer is ideally kept stationary in the cold end while the piston moves outward under the persuasion of the higher internal pressure. This is the expansion stroke of the cycle. Now at the beginning of the expansion stroke, most of the engine air was at the highest temperature, namely that of the hot end. As the expansion proceeds, some air moves into the cold cylinder, so the average temperature drops, but this average is still always higher than the cold end temperature. For simplicity of discussion assume the expansion continues until the pressure inside has dropped exactly back down to atmospheric.

In the next part of the cycle, Fig. 16c, the air in the hot section is transferred to the cold by moving the displacer. For maximum effect, the piston is nearly stationary during this process. The end result is a constant volume cooling of the engine gas. This drops the pressure below atmospheric. Again the energy for this stroke is small and is supplied by a flywheel through linkage.

In the final step of the cycle, Fig. 16d, the piston moves inward under the influence of higher pressure outside. As the gas volume is compressed, its pressure rises until at the end it equals atmospheric. This compression returns the device to the initial state from which the process of Fig 16a began. The 4-step cycle just described is a simple form of the Stirling cycle, named after Robert Stirling, a devout Scottish minister who invented the engine based on the cycle in 1816.

A Complete Stirling Engine

It should be clear that with the sequence of movements described above, useful work will be produced each cycle. However the piston and displacer will not just move that way all by themselves. Some machinery is needed for that. Figure 17 shows the piston-displacer unit of Fig. 16 linked to two ordinary crankshafts. The cranks are connected through a chain and sprocket and so they rotate

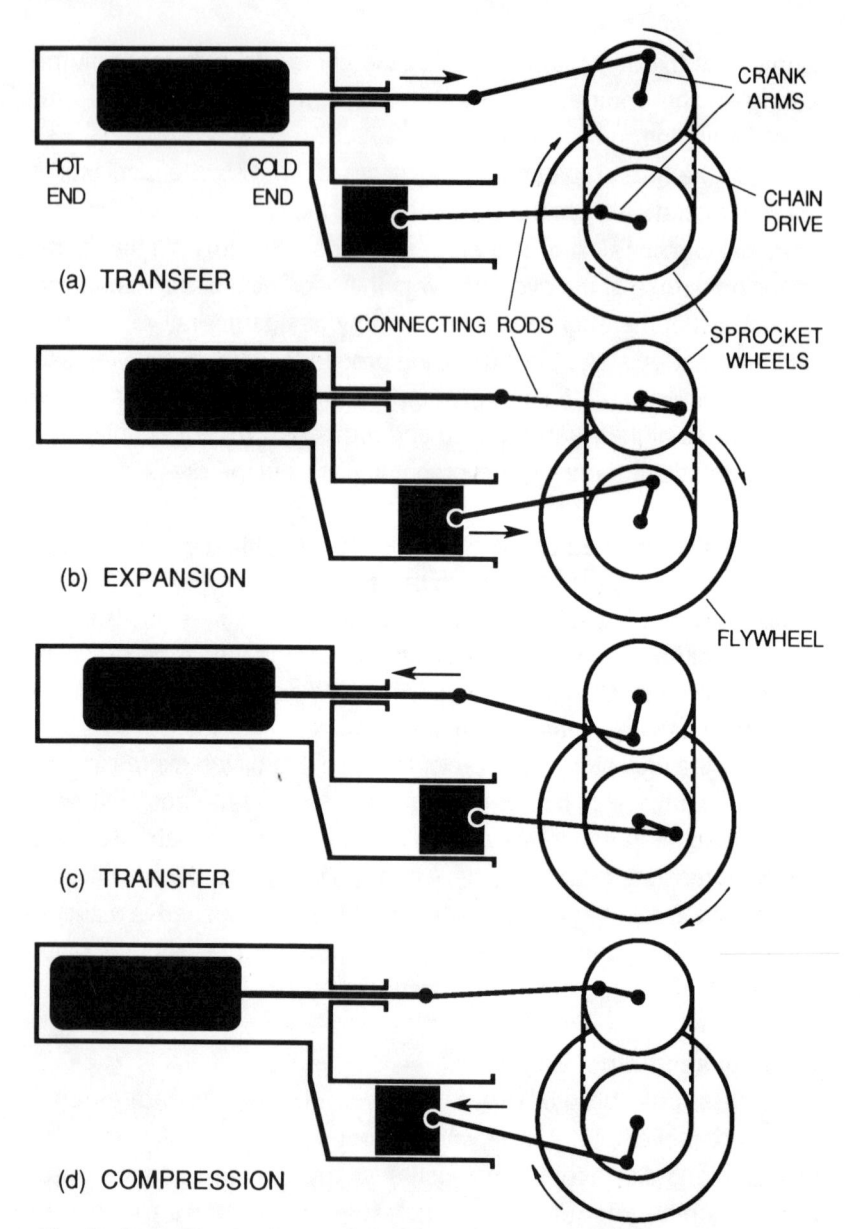

Fig. 17 Connecting the displacer rod and piston to cranks which are 90° out of phase automates the cycle of Fig. 16 to make a version of Stirling engine known as the "split-cylinder" type. A flywheel is necessary to sustain operation during the transfer strokes.

together in synchronization. They are set so that the crank driving the displacer is about 90° ahead of the one linked to the piston.

In Fig. 17a, the displacer crank is near its midstroke position, so the displacer is moving at its peak speed from the hot end to the cold. It is rapidly transferring air from the cold to the hot section. Meanwhile, the piston is not moving much at all since its crank is close to its top dead center. A quarter turn later, in Fig. 17b, the piston is moving outward rapidly on the expansion stroke while the displacer crank is rounding a dead center. Another quarter turn finds the piston all the way out and the displacer rapidly moving the air from the hot to the cold end. In the last 90° of rotation, the piston moves inward on the compression stroke while the displacer dwells in the hot end.

With the mechanical drive system described, in each revolution of the cranks, the four processes of heating, expansion, cooling , and compression are carried out with mechanical regularity and in rapid succession. Now they are not carried out perfectly from the thermodynamic point of view. This is because neither the piston nor the displacer are absolutely motionless for any length of time during which only the other moves. This makes the cycle a little less effective than the ideal cycle described with reference to Fig. 16. It is however a reasonably good approximation and is one that can be performed very rapidly with high mechanical efficiency.

Other Engine Mechanisms

An incredible variety of mechanisms have been devised for Stirling cycle engines. Most mechanical drives for the split-cylinder type Stirling engine are based on the ordinary crank and connecting rod. The simplest system is a single shaft which carries both cranks. This arrangement has been widely used on small model and toy hot air engines. Some larger Stirling engines intended for practical tasks used the same system, such as the well known Ky-Ko fan. These fans had a 20 inch diameter blade which turned at 360 rpm and ran for 10 hours on a pint of kerosene . Although manufactured in England, many Ky-Ko fans made their way throughout the empire to tropical regions of the world. One of these fans can be seen in the 1938 movie classic "The Beachcomber" with Charles Laughton and Elsa Lanchester. In one all too brief scene a Ky-Ko is running on a night stand providing some measure of comfort for a bed ridden missionary. A fan originating in India with a similar mechanical arrangement is featured in the MGM movie "Lady of the Tropics".

A simpler mechanical system is possible with a "V" arrangement of the cylinders. This allows a single crank to drive both the piston and displacer. This is about as simple a mechanism as one could hope for, but a disadvantage of this configuration is the rather long connecting passage necessary to connect the cylinders.

One of the most elegant mechanisms is the "automatic" one invented by Ossian Ringbom in 1907. His idea did away with just about half of the drive machinery. The displacer was fitted with a fat displacer rod. It is big enough to act as a secondary piston to push and pull the displacer back and forth. With everything proportioned just right, the displacer moves in the correct way to run the engine, almost as if by magic. A properly designed and well made Ringbom engine is quite a sight to see in operation!

Stirling engines capable of operating on small temperature differences are a recent development. Ivo Kolin of the University of Zagreb exhibited a small engine in 1983 that would operate with just 15°C (27°F) differential in temperature between the source and sink. At the Univ. of Wisconsin-R.F., Ringbom engines have been developed that operate down to a 5°C difference. Engines of this kind of necessity have a different shape than the conventional Stirling engines which run at differences up to 600°C. Low temperature difference Stirling engines have large diameter and short displacer chambers, and a low compression ratio. Their power is low compared to their size, but they make excellent and amusing engines from which to learn first principles. The latest in the line of low temperature differential Stirling engines is the "P-19" developed at the University of Wisconsin-R.F. which will run on a difference as small as 1/2 °C (less than 1°F) . It runs if you just pick it up and rest it on your hand !

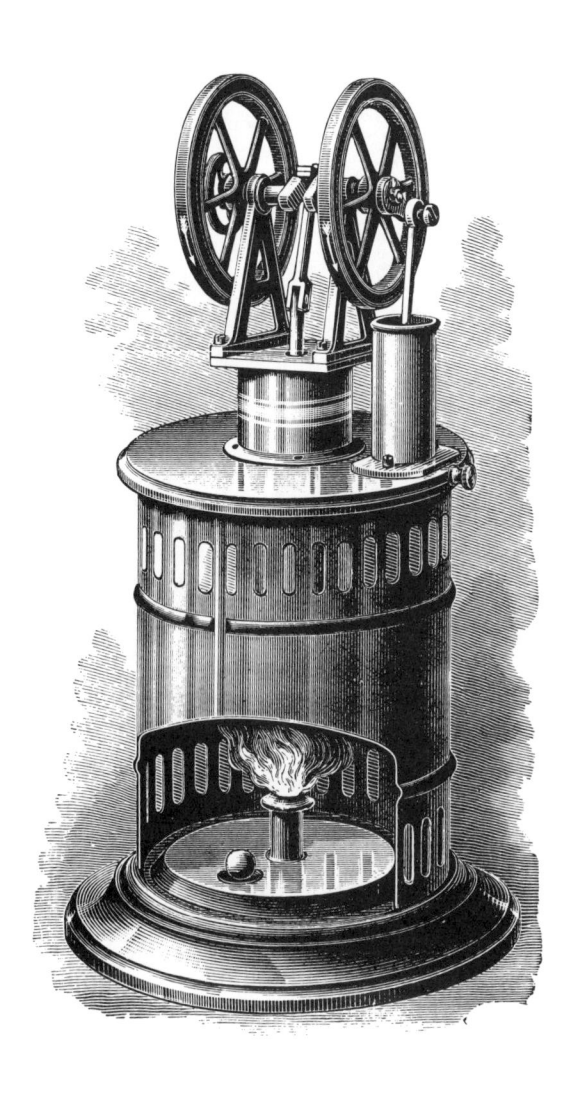

A model hot air engine of the split-cylinder type dating back to about 1900. Both cranks are carried on a single shaft and are set 90° apart. A drilled hole in the plate forms the connecting passage between the piston cylinder and the displacer cylinder.

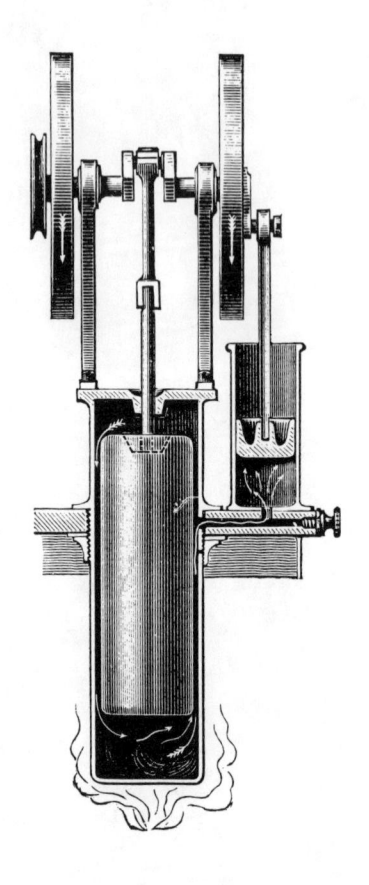

A cutaway view of the model engine shown in the preceding figure. The piston is 0.66" (1.67cm) in diameter with a stroke of 0.875" (2.22cm). The displacer diameter is 1.25" (3.17cm) and the stroke is also 0.875" (2.22cm).

The "KY-KO"
NON-ELECTRIC FAN

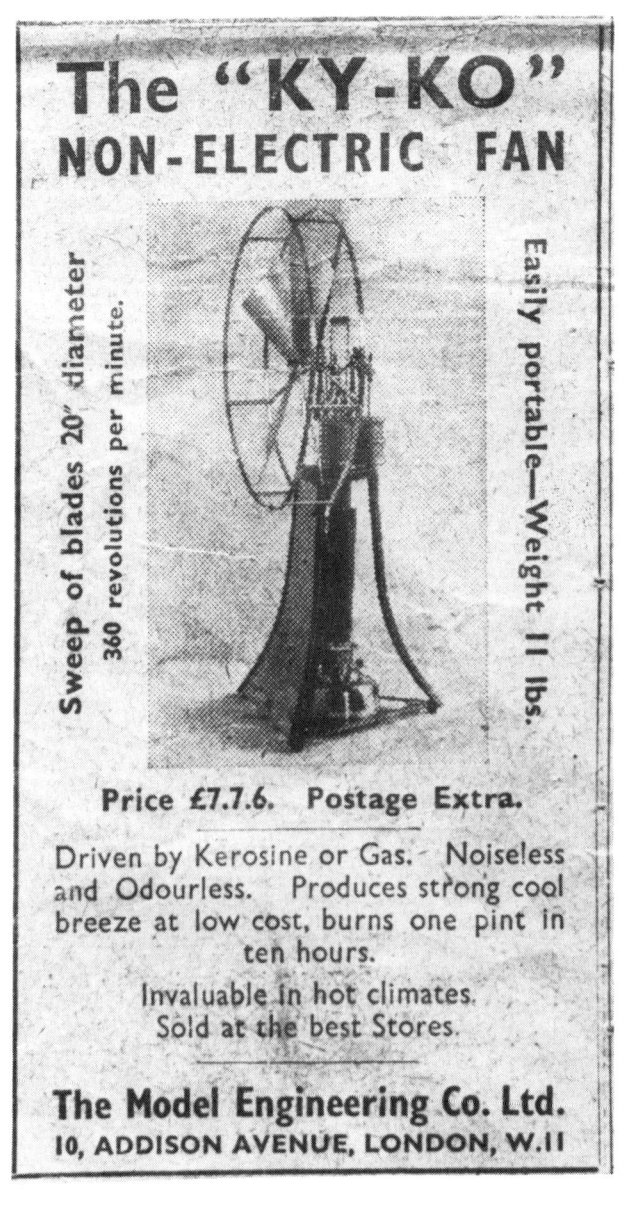

Sweep of blades 20" diameter. 360 revolutions per minute.

Easily portable—Weight 11 lbs.

Price £7.7.6. Postage Extra.

Driven by Kerosine or Gas. Noiseless and Odourless. Produces strong cool breeze at low cost, burns one pint in ten hours.

Invaluable in hot climates.
Sold at the best Stores.

The Model Engineering Co. Ltd.
10, ADDISON AVENUE, LONDON, W.11

An advertisement for the Ky-Ko fan which appeared in the London Times in 1938. Fans similar to the Ky-Ko are still being manufactured in Pakistan and feature ball bearings on the main shaft.

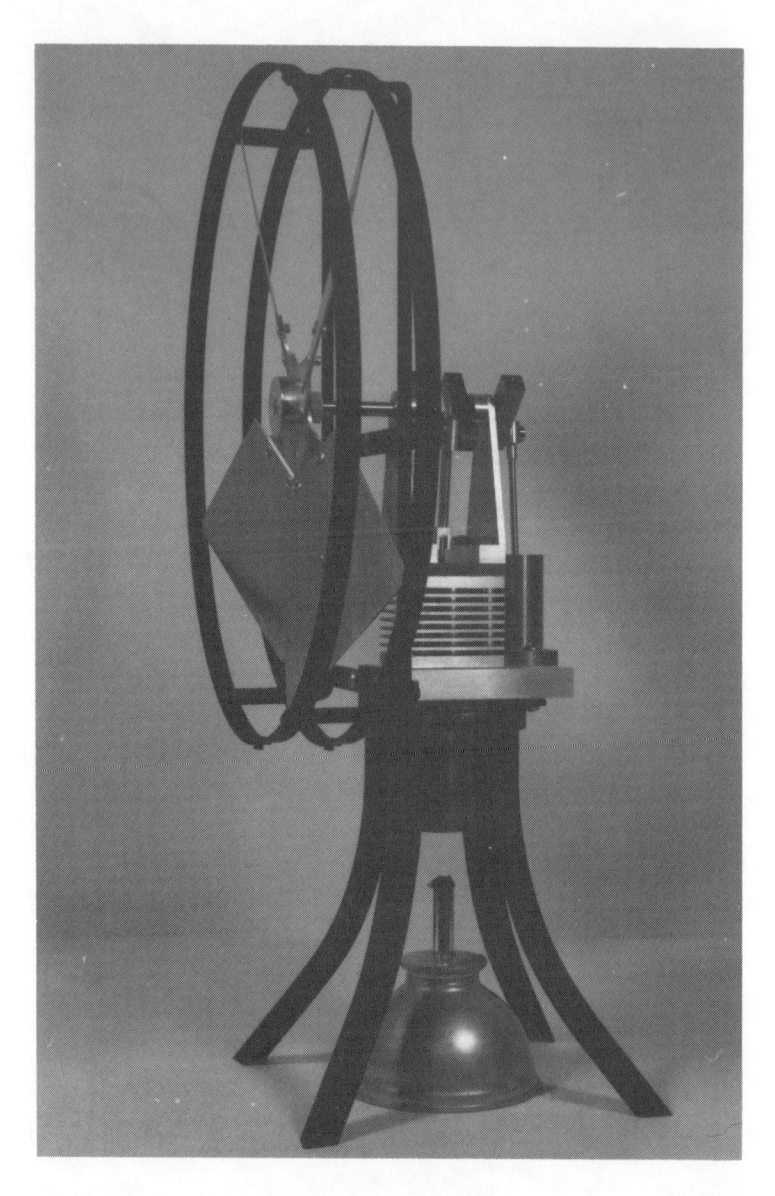

"Moriya", a home-made Stirling powered fan. Heated by an alcohol flame, the engine spins its 10" blade at about 900 rpm and provides a pleasant gentle cooling breeze.

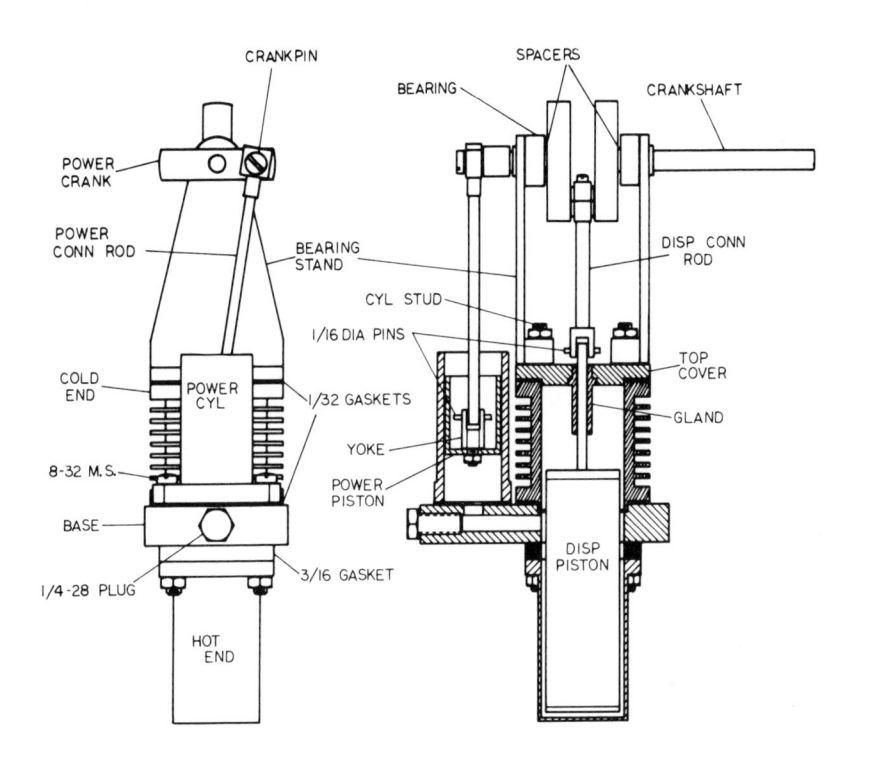

A scale drawing of the split-cylinder type engine of Moriya. Piston diameter is 3/4" (1.90cm) and the displacer diameter is 1" (2.54cm). Each has a stroke of 1". The engine can be built from scratch using a lathe, drill press, and hand tools.

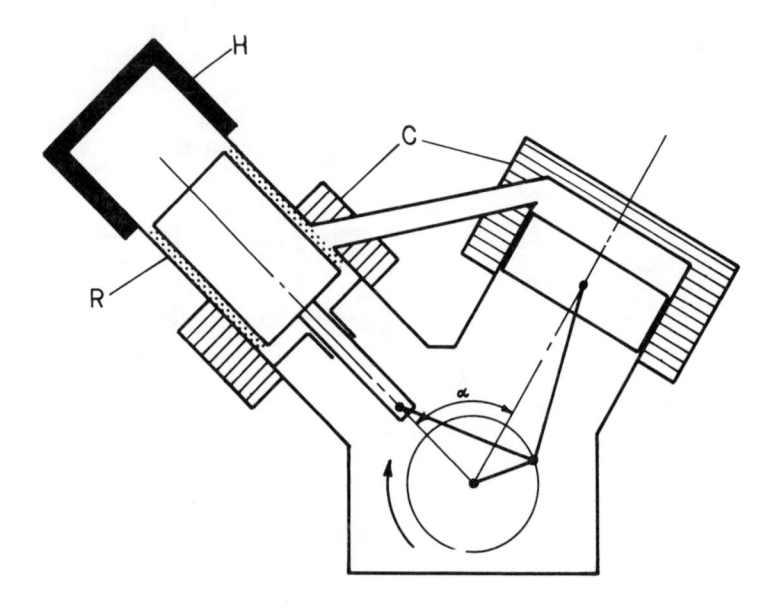

A "V" arrangement of the piston and displacer cylinders allows a single crank drive for the split-cylinder Stirling engine.

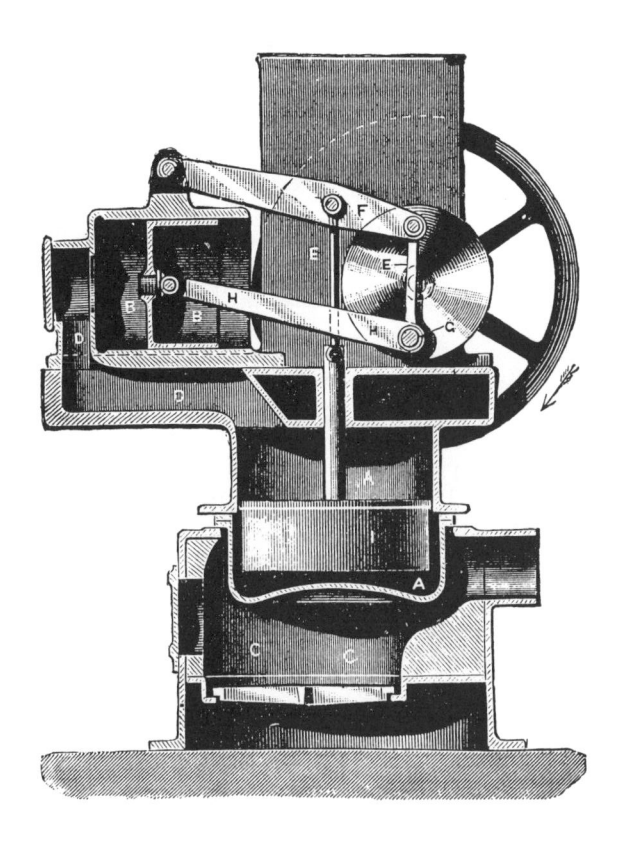

The compact Robinson engine from the turn of the century featured a clever linkage drive. These engines were made in assorted small sizes for domestic use such as driving sewing machines, pumps, and fans, and used coal, wood, or gas as fuel.

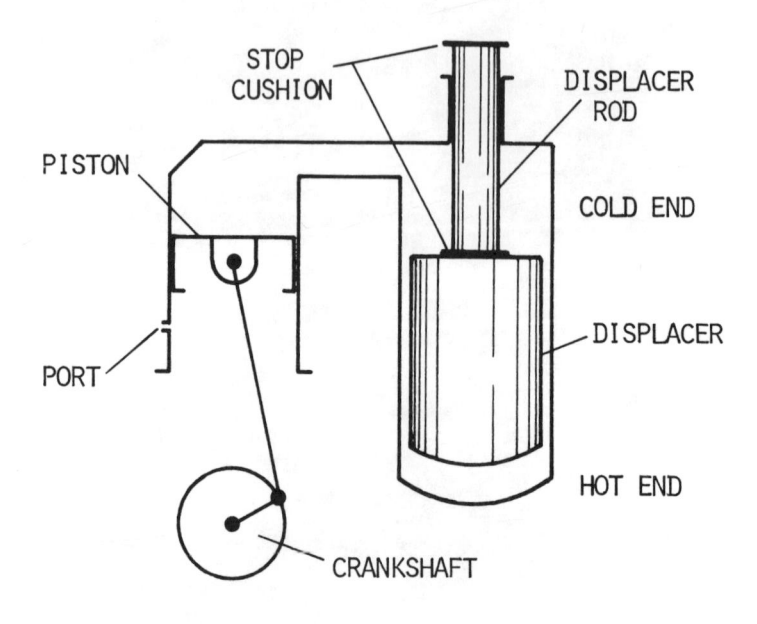

The Ringbom engine has no mechanical drive to the displacer at all. The displacer is instead driven by the changing pressure within the engine. The rod on the displacer is made large enough in diameter to respond to these pressure changes. Modern Ringbom engines are capable of high speed stable operation and offer mechanical simplicity and great flexibility in design.

A model Ringbom engine nicknamed "Tapper" is shown here actually running on a tiny alcohol flame (not visible in the photo) as the heat source. This 2cc engine is capable of speeds over 1000 rpm and operates with as regular a beat as any Stirling engine having a full mechanical drive system.

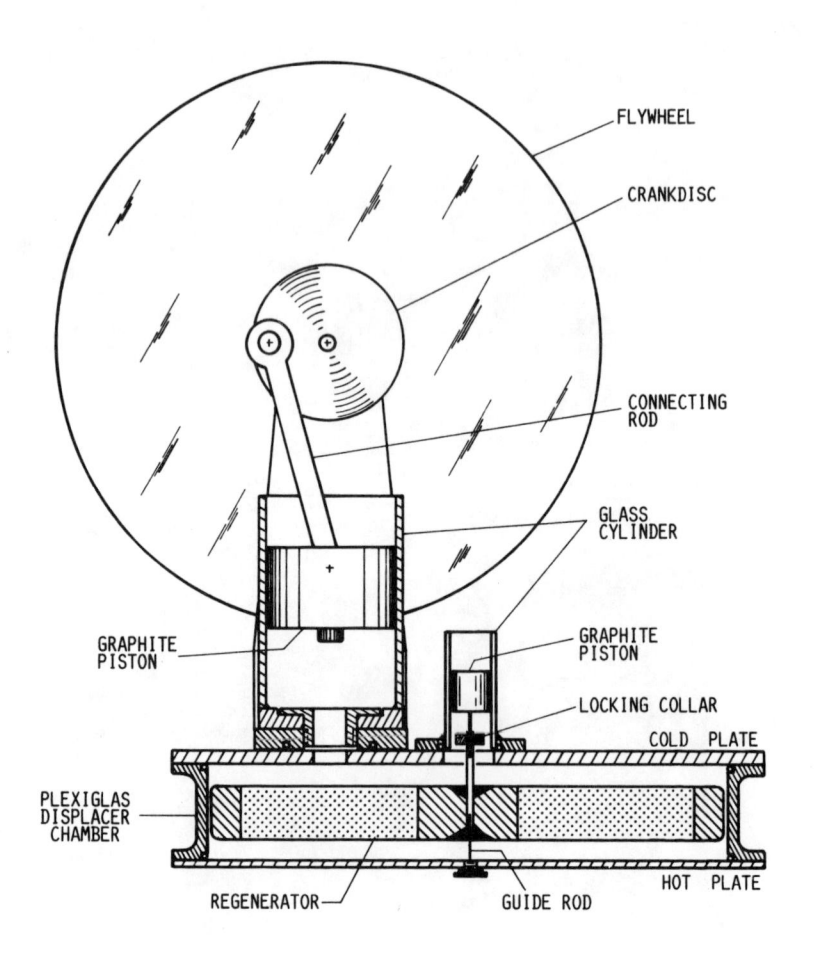

A sectional drawing of a small Ringbom engine, the "L-27", designed for operation on temperature differences between source and sink of from 5 to 100°C. The main piston has a swept volume of 25cc and is connected to a crankshaft in the usual way. The displacer is 5.1" (13 cm) in diameter and is driven by a tiny attached piston. The engine runs for about two hours when placed on top of an insulated container containing a few cups of heated water.

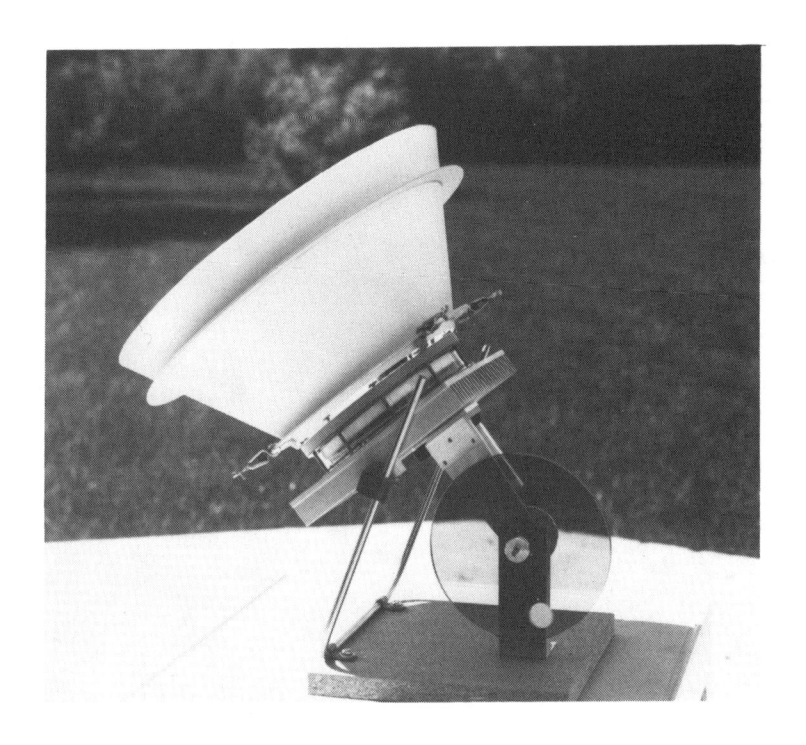

The characteristic large diameter displacer of low temperature differential engines makes them ideally suited for operation on passive solar energy. This photo shows the L-27 Ringbom engine setup for solar operation with a conical reflector and a finned cooler as part of an energy research project sponsored by the Charles A. Lindbergh Foundation.

The "P-19" Stirling engine was designed and built in 1990 specifically to operate with ultra small temperature differences between its warm and cool sides. The engine runs at over 100 rpm when just held on a human hand. With a cold drink resting on its top plate, P-19 will run on for hours.

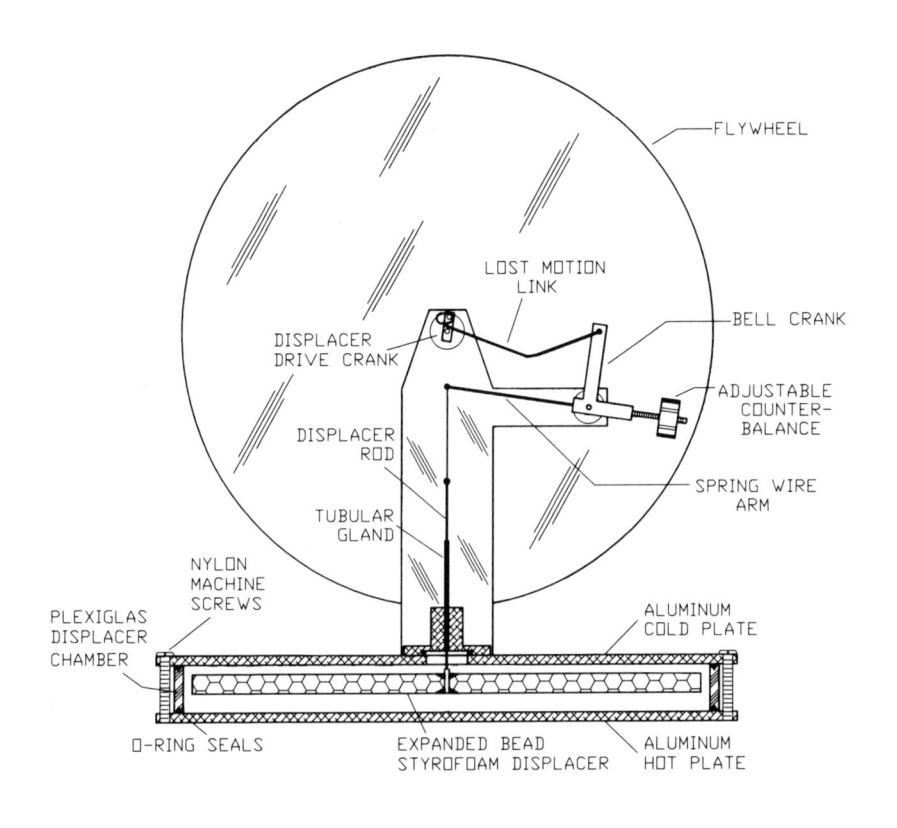

The P-19 Stirling engine has a compression ratio of just 1.004 to 1 and has operated with only 1/2°C (less than 1°F) difference in temperature between the two end plates of the displacer chamber. The balanced displacer drive system has a "lost motion" link which allows the displacer to dwell at the ends of its stroke, giving more time for the expansion and compression processes.

Heat Losses

As discussed above, the displacer is an ingenious device which makes rapid operation of the Stirling engine possible while improving efficiency. The displacer removed the necessity of repeatedly heating and then cooling the same cylinder material just in order to heat and cool the working air within. Only the heating and cooling of the air inside produces work. Cyclically heating and then cooling the cylinder material just transports thermal energy from the source to the sink. No work is produced from this heat; its potential to do work is just wasted.

This thermal short circuit is greatly reduced by the displacer. The engine gas is alternately heated and cooled by being transferred by the displacer between the hot and cold sections of the engine. The material walls of these sections are maintained at the same temperature during the operation of the engine. The hot end is continually kept hot. The only heat that needs to be supplied is that to make up for what the engine air takes from it each cycle when it is shuttled into the hot end and expands. So the heat added has nothing to do with the cylinder material, but only with the working air inside.

Similar remarks apply to the heat removed from the cold end. The engine material in this region is kept at a constant cool temperature. The only heat removed from it is what the air inside gives to it when it is transferred into the cold section and compressed. Thus the effect of the displacer on efficiency is to avoid heating and cooling engine parts without that heat passing into and out of the working substance.

Of course there are some imperfections in practice. With one end hot and the other cold, some heat travels by conduction through the cylinder material direct from the hot end to the cold. This is a loss in efficiency, but it can be minimized and held to acceptable limits by making the walls of the displacer chamber as thin as practical, especially in the middle of this section. In some small Stirling engines, the wall thickness of this section is as little as .005". In flame heated engines, stainless steel is the favored material for the displacer and its cylinder. Stainless steel resists oxidation and has a lower conductivity than other common metals. The lower conductivity is not a real obstacle to passing thermal energy in and out of the working air since the wall is relatively thin, but does substantially reduce conduction loss from the hot to the cold end where the thermal path is much longer. The larger the temperature difference between the hot and cold ends, the longer should the cylinder be to reduce the conduction loss.

Ideally, the center section should be an insulator, but this is difficult to accomplish in practice for high temperature engines. One end might well be red hot and the other near room temperature. The only suitable insulating material for such extremes would be a ceramic which has yet to be successful in withstanding the rigors of real usage. On engines operating from lower temperature sources, such as from warm water or passive solar energy, plastic can be used for the midsection. Acrylic is a very good insulator and so the displacer section can be made quite short in such engines. In summary, an obvious requirement for an efficient Stirling engine is minimal material conduction paths from the hot side to the cold. In addition, there is an important internal device called a regenerator that can drastically improve the efficiency of a Stirling engine.

The Regenerator

In addition to thermal short circuits through the structure of a Stirling engine, there are internal thermal losses. The principal loss arises from the cyclic heating and cooling of the engine gas itself. Now cyclic heating and cooling is necessary to do the expansion and compression that ultimately produce the work output of the engine. But the way the heating and cooling is carried out can heavily influence the efficiency of the engine.

In an engine with a plain displacer, the heating and cooling can be improved because not all contributes to the work done by the engine. Some of the heat supplied to the engine goes to just raising the temperature of the air as it enters the hot space before it expands to do work. More heat is added then to do the expansion. After expansion, the warm gas is transferred to the cold end where heat is given up through the cold engine parts to the sink before the compression. Thus looking carefully, one can see heat transported through the engine working medium from the source to the sink that doesn't directly contribute to the work produced. It is thermal energy that is absorbed from the source during a constant volume process when no work is being done and is rejected to the sink also during a non-work producing process. This energy flow through the engine can now be easily spotted in Fig. 16; what heat goes inside the engine in Fig. 16a comes right back out in Fig. 16c. A flow of heat like this is a lost opportunity to produce work and so is a loss in efficiency.

Along with his engine, Robert Stirling invented a device to counter this loss. He called it the "economizer", but is now usually called the regenerator. In the first Stirling engine the regenerator was a matrix of relatively fine wire attached to the outside of the displacer body. The wire was wound around the displacer with straight wire spacers to produce a basket type weave. The matrix just about filled up the annular gap between the displacer and its cylinder, but some clearance was necessary to prevent rubbing; rollers in the midsection kept the displacer centered in the bore. The weave of the regenerator was open enough to encourage the displaced air to flow through it freely.

To understand the operation of the regenerator, consider the split-cylinder engine that was illustrated in Fig. 16, but now with a regenerator in the annular space around the displacer. Figure 18 shows the added regenerator. Imagine that the engine is all warmed up and has been running for a while. The hot end is hot and the cold end is relatively cold. Inside, the displacer is also hot at one end and cold at the other. So is the regenerator matrix.

After the expansion stroke, the air in the displacer chamber is hot. As the displacer moves into the hot end, this hot air is displaced and flows through the regenerator on its way to the cold end, Fig. 18c. As it passes around the many wires it gives up a little of its thermal energy to each strand. The air thus gradually drops in temperature as

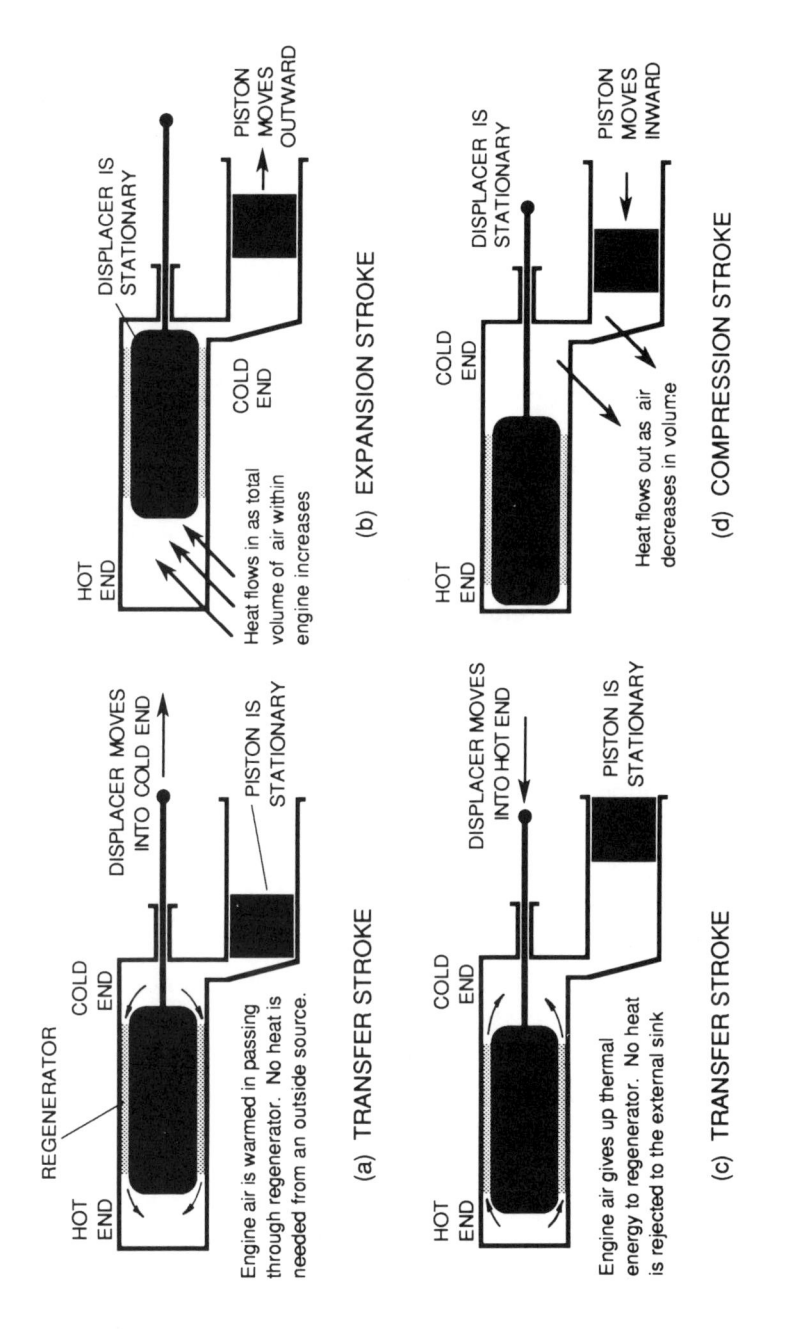

Fig. 18 The operation of the split-cylinder type Stirling engine equipped with a regenerator.

it moves along through the regenerator. By the time it exits, it is already just about down to the temperature of the cold section of the engine. Contrast this with the corresponding process in the engine without the regenerator. In Fig. 16c, heat is rejected to the outside, but in Fig. 18c, this thermal energy is captured and held within the engine in its regenerator.

During compression, Fig. 18d, heat flows out of the cold section of the engine, just as in Fig. 16d. This step is identical in engines with or without a regenerator. Now in the next transfer stroke, Fig. 18a, the displacer moves into the cold end and the cold air flows back into the regenerator. As it passes through, it picks up the thermal energy residing in the matrix wires and warms up again. When it exits, it is just about back up to the hot end temperature. Without a regenerator, Fig. 16a, heat must be supplied from the outside in this stroke. This is exactly how the regenerator improves efficiency. The expansion stroke which follows, Fig. 18b, is the same as before, requiring just the same amount of heat as the engine without a regenerator.

The operation of the regenerator is like that of the heavy woolen scarf a mother wraps over the face of her child going out to play in the winter cold. It is not there just to keep the child's face warm from the eyes down, but it also acts regeneratively to keep the respiratory system warmer internally. Warm exhaled air passes through the weave of the scarf and leaves thermal energy in the fibers. When cold fresh air is inhaled, it goes through the same area of the scarf, picks up the thermal energy, and enters the nose already warmed up. The fibrous structure of the wool is very fine so a lot of surface area is presented to the air passing through so it can rapidly regenerate. It is amazing how technically advanced mothers are at caring for their children; for ages they were way ahead of Stirling engine technology! In an engine, the regenerator picks up thermal energy from the hot gas on its way to the cold section, stores it, and then half a cycle later returns it to the cold gas on its way back to the hot end.

If a regenerator worked perfectly, only whatever energy is needed for the expansion would have to be taken in from the source. This is clear from just looking at the cycle steps in Fig. 18. In practice, regenerators do not work perfectly, but it is not too hard to make them work quite well. Even a poor regenerator improves efficiency to some extent. Regenerators also increase power by making higher speeds possible. Less thermal energy needs to be absorbed in each cycle when a regenerator is working, so each cycle simply takes less time.

Regenerators have taken many forms. Some are compressed metal wool like scouring pads. The Robinson hot air engines made in England around 1900 for light domestic chores featured a hollow displacer having perforated ends and filled with metal wool. Very effective regenerators can be made of layers of screens. In low temperature differential Stirling engines, most of which are used for demonstration and experimentation and not for practical work, regenerators are important to give higher speeds and longer runs from limited heat sources such as a cup of hot water. The model engines made by the New Machine Company have an integral regenerator-displacer made entirely of porous foam.

Another form of regenerator is a parallel array of thin metal plates spaced a small distance apart. This also presents a large surface area to the air flowing through the spaces. Regenerators can also be fitted into engines so as to be stationary. This is a particular advantage for those engines intended for high speed operation, since it reduces the reciprocating mass. In such engines, the regenerator is housed in a flow circuit around the cylinder in which the displacer reciprocates. The displacer then causes the engine air to flow back and forth through the circuit containing the regenerator. This circuit can then also contain a finned or tubular heater and cooler section to enhance the heat transfer into and out of the engine. This is the form which modern high speed Stirling engines follow.

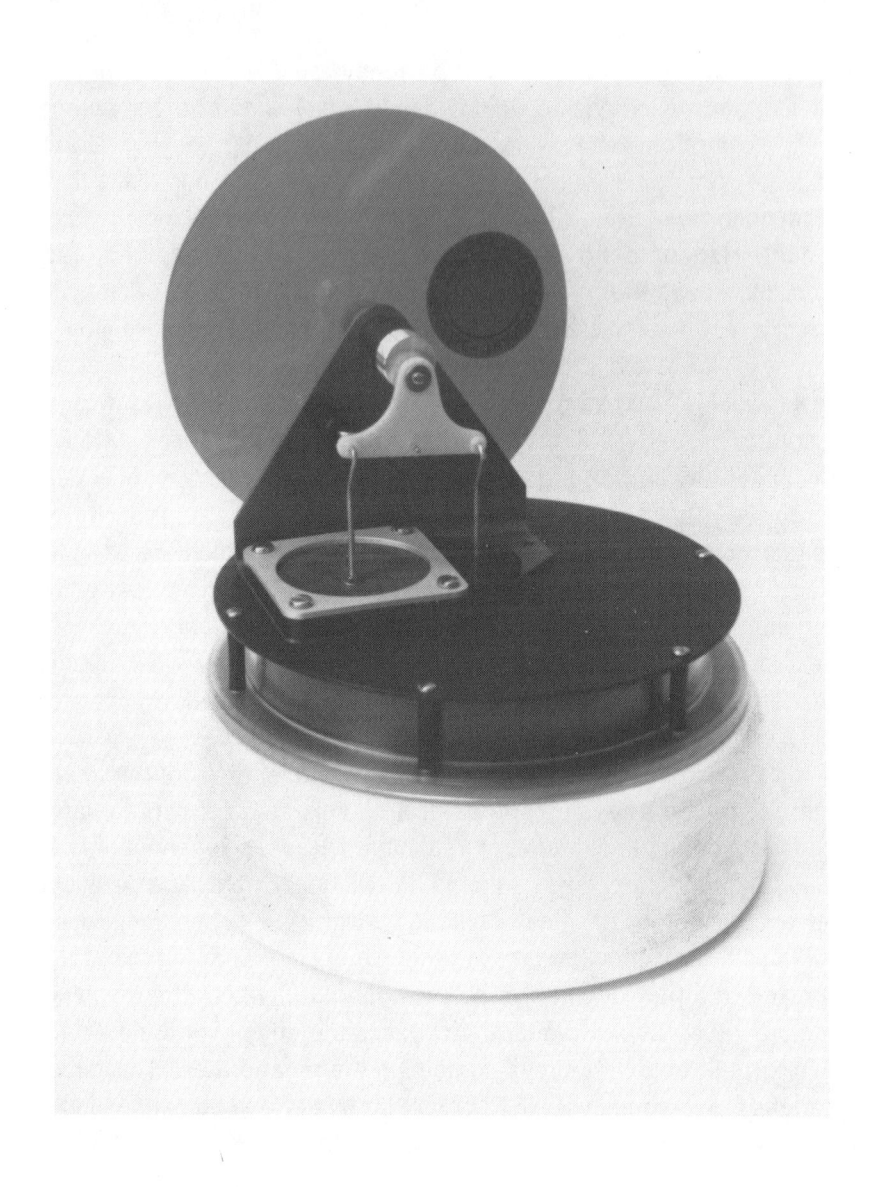

A model engine made by the New Machine Company for operation on warm water features a displacer entirely made of porous foam which acts as a large regenerator.

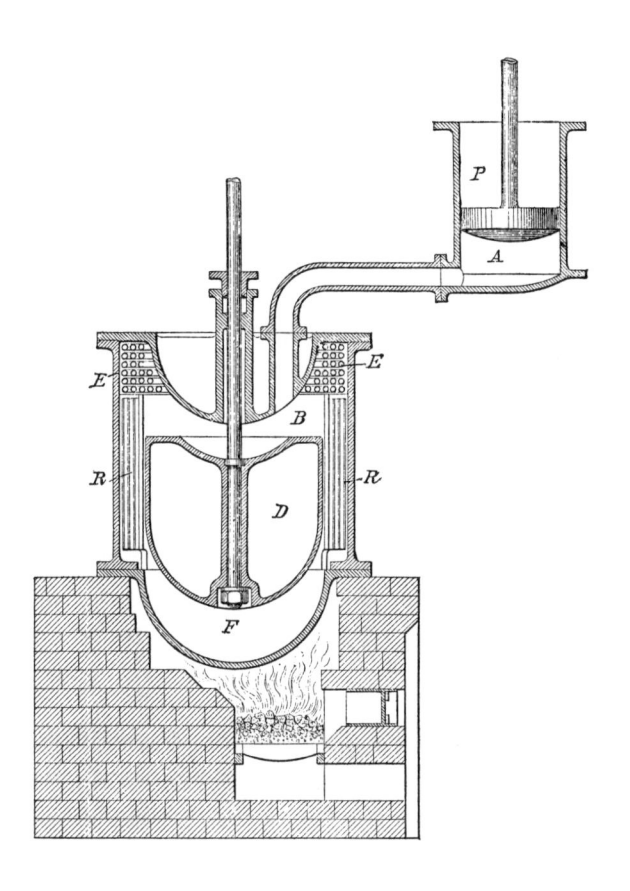

Schematic drawing of an early Stirling engine having a stationary regenerator. The rengenerator is housed in an annular space surrounding an inner cylinder in which the displacer works. The displacer fits the inner cylinder fairly close so that as it reciprocates, the engine air must travel around the cylinder and pass through the regenerator space. In this case the regenerator is a round array of thin metal plates spaced a small distance apart.

The Single-Cylinder Configuration

In the engine described in Robert Stirling's first patent, the piston and the displacer shared a common cylinder. In that engine the power cylinder was not separate from the displacer cylinder, but rather was simply an extension of it. This arrangement, now known as the "single-cylinder" or "beta" configuration, makes it more complicated to arrange the mechanical driving system, but it has the potential thermodynamic advantage of allowing more of the expansion to take place in the hot section at a higher pressure. One of the complications is the necessity of having the displacer rod pass through the piston for connection to the linkage since a close sliding fit would be impossible to maintain through the red hot end of the engine.

The connecting passage of the split-cylinder configuration described above represents an unused section of the engine. Neither the piston nor the displacer sweeps through this "dead" space. Dead space anywhere inside an engine degrades its power output by diminishing the pressure variation over the engine cycle. Because the single-cylinder arrangement does not require a connecting passage, dead space is reduced. Every bit of space in the main cylinder, except for the clearances, is utilized by the piston or the displacer. Indeed, some sections of the engine are swept by both. This can be seen in Fig. 19 which shows the four steps of the single-cylinder Stirling cycle.

In Fig. 19a, the displacer is moving toward the piston in the cold section. This transfers the air into the hot end and raises its temperature and pressure in the usual way. The displacer moves all the way to the piston and so all the air ends up in the hot end. The expansion stroke follows. As shown in Fig. 19b, the piston and displacer move out together all during the expansion. This keeps the air completely in the hot section and maintains the temperature and pressure as high as possible all through the expansion until the piston reaches it outermost limit.

In the cooling step, the engine air is transferred into the cold section by the movement of the displacer into the hot end as shown in

(a) TRANSFER STROKE

DISPLACER MOVES INTO COLD END UP TO PISTON

HOT END   COLD END   PISTON IS STATIONARY

(b) EXPANSION STROKE

PISTON AND DISPLACER MOVE OUTWARD TOGETHER

HOT END   COLD END

(c) TRANSFER STROKE

DISPLACER MOVES INTO HOT END

HOT END   COLD END   PISTON IS STATIONARY

(d) COMPRESSION STROKE

DISPLACER IS STATIONARY IN HOT END

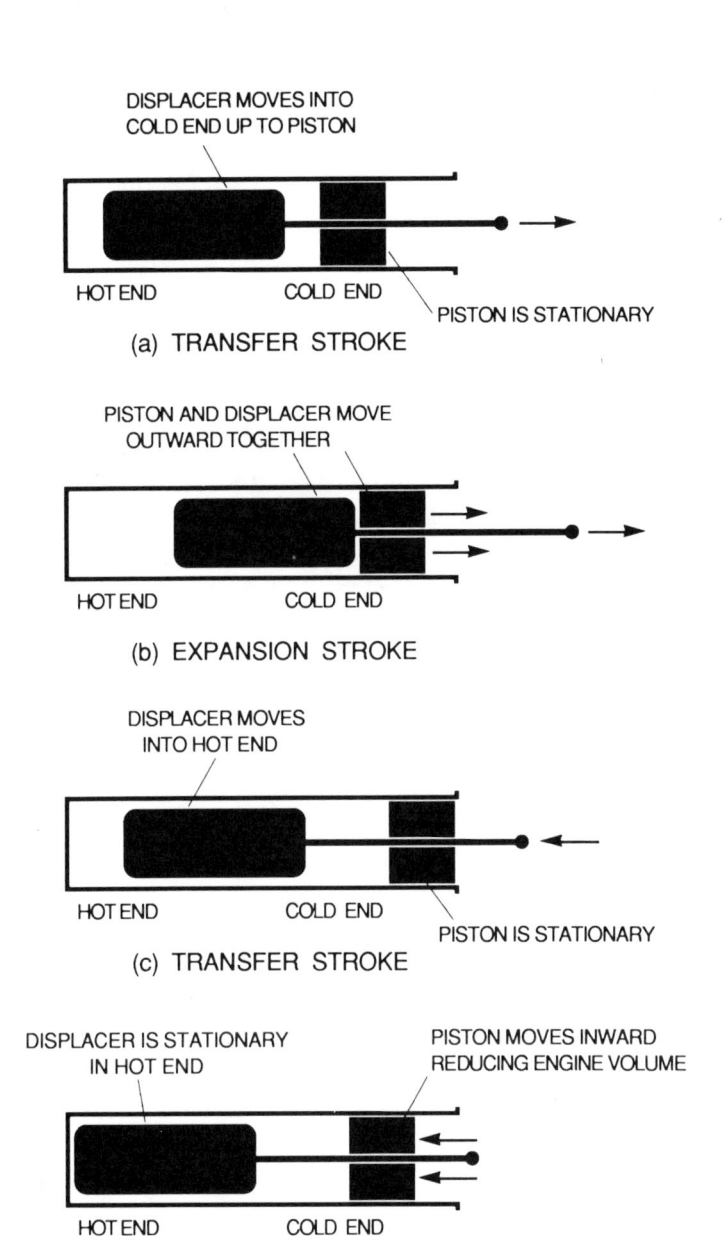

PISTON MOVES INWARD REDUCING ENGINE VOLUME

HOT END   COLD END

Fig. 19 The four steps of the ideal Stirling cycle with piston and displacer in a single cylinder.

Fig. 19c. The piston is ideally stationary during this transfer stroke also. The temperature and pressure of the air fall and then the piston moves inward on the compression stroke, Fig. 19d. After reducing the volume to about half, the cycle is ready to be repeated.

The four steps for the single-cylinder Stirling cycle are about the same as for the split-cylinder type of Fig. 16. The only difference is in the expansion stroke, but this is an important difference. In the single-cylinder engine, the expansion all takes place in the hot space. In the split-cylinder type engine, it is divided between the hot and cold sections so pressure and work output are somewhat less.

Although the power output is better for the single-cylinder engine than for the split-cylinder type, the mechanical drive system is harder to make. Most of the early single-cylinder Stirling engines have quite ingenious linkage drive systems. These mechanisms tended to be large and complex, but they worked well on the slow moving hot air engines of the early days.

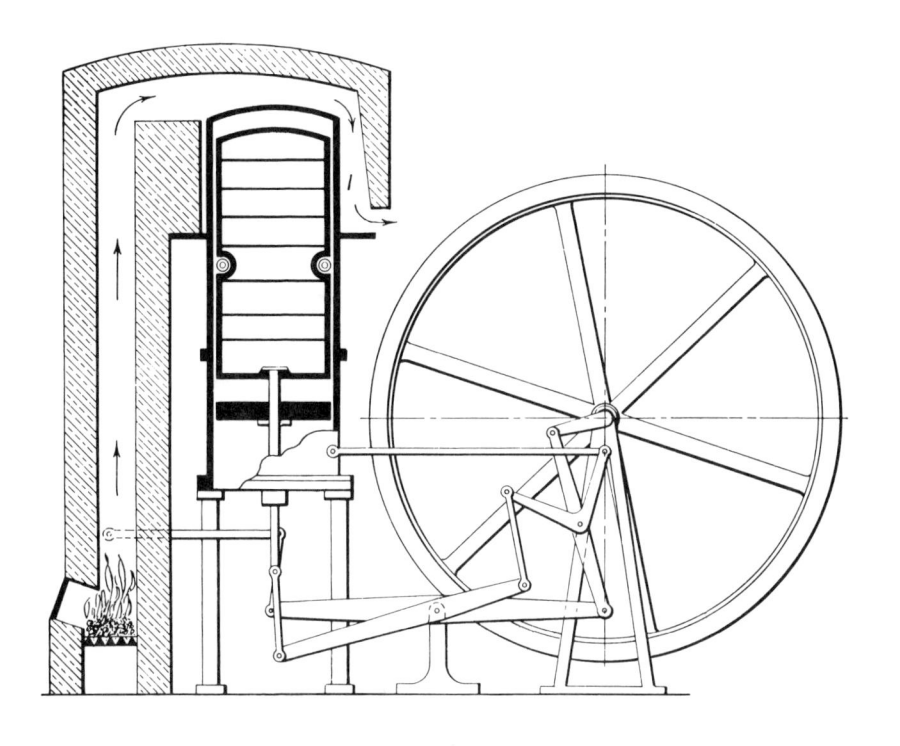

A drawing of the first Stirling engine which was invented in 1816. The main cylinder was about 6.5 ft (2m) long and 2 ft (0.6m) in diameter. The engine was heated by coal and produced about 2Hp (1500 W). It was put to work pumping water from a quarry which it did very efficiently compared to the steam engines of the day. After several years of duty, the hot end was damaged through accidental overheating. (illustration courtesy of T. Finkelstein)

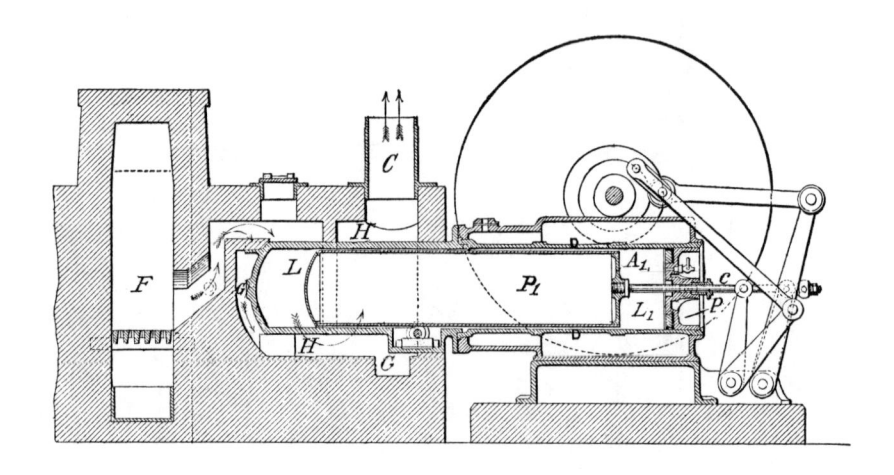

A cutaway drawing of the Bailey hot air engine from about 1860. This engine had a cylinder bore of 14.6" (37cm) and a piston stroke of 6.9" (17.5cm). It operated at just over 100 rpm and delivered 1.3 Hp (0.97 kW) out of the shaft. Note the roller supporting the long displacer.

An interesting linkage drive was used on the Stirling engine designed by John Ericsson in 1880. A crank, connecting rod, lever, and drop link drove the piston. The same crankpin drove the displacer through a second connecting rod, a bellcrank, and a long U-shaped link straddling the main cylinder.

The Ericsson engines were manufactured in a range of sizes suitable for pumping water to country houses, resorts, greenhouses, hospitals, factories, and farms. Cylinder bore sizes were available from 5 to 12 inches and the engines could be had with furnaces for burning coal, wood, or gas. In operation, the lower end glowed a dull red above the fire while the opposite end was kept cool as the water that the engine pumped flowed through a jacket surrounding the upper end before being discharged.

In a typical application, the engine would draw water up a pipeline from a well and pump it up to a supply tank above the establishment it served. Gravity then fed water throughout the building. The engine would only have to be run to refill the tank whenever necessary. A wood or coal fire would be started and when hot enough, just one turn of the flywheel would start the engine pumping silently away. With this, one could walk away and the engine would continue to run for an hour or two until the fire died out.

The 8 inch (20 cm) bore Ericsson engine could pump 500 gallons (1900 liters) of water per hour to a height of 50 feet (15.2 m). The engine was about 5-1/2 feet (1.7 m) high, weighed 700 lbs (320 kg), ran at 80 rpm, and used 3-1/2 lbs (1.6kg) of coal per hour. The Ericsson hot air engines were very popular and highly regarded for water pumping. They were steadily improved and remained in production until about 1920. Many Ericssons have been preserved and can occasionally be seen running at farm threshing shows.

Many other drive systems for single-cylinder Stirling engines have been devised. Modern versions are compact and suitable for high speed operation.

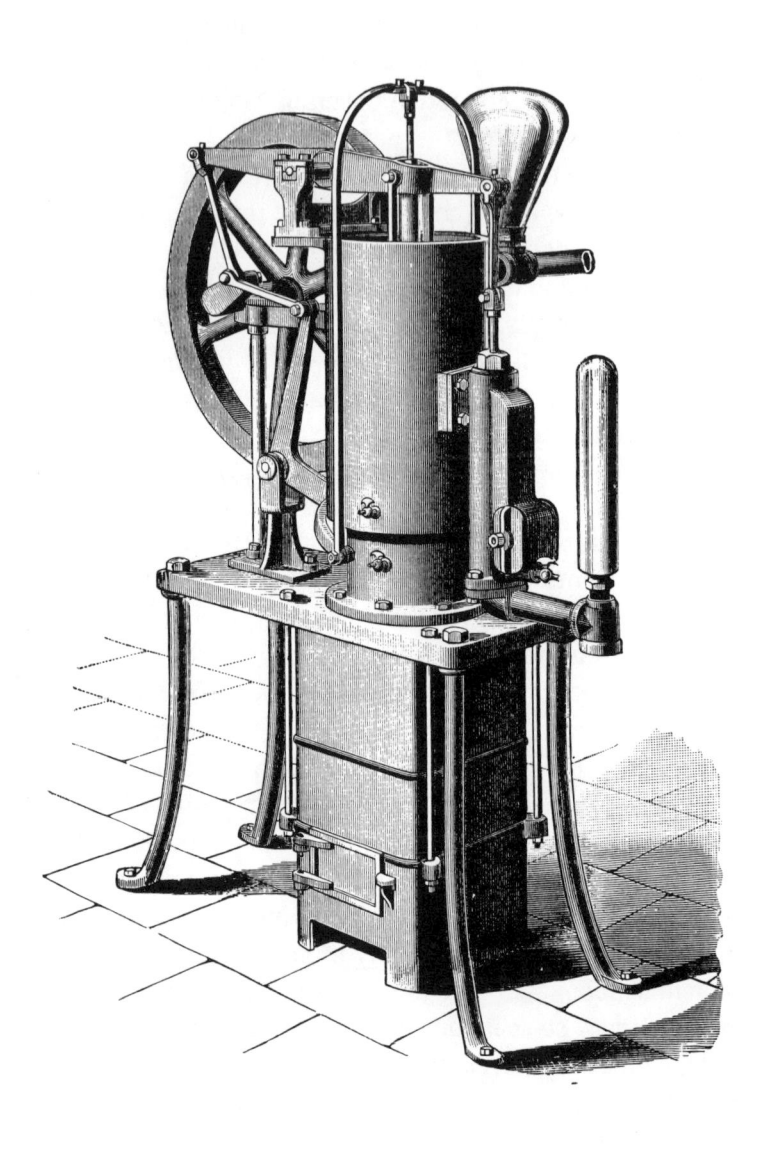

An Ericsson hot air water pumping engine as depicted in a catalog from 1890 (illustration courtesy of Alan Phillips).

1. Cylinder.	10. Beam Centre Bearing.	19. Pump Chamber.	28. Gas Cock.
2. Air Piston.	11. Connecting Rod.	20. Pump Gland.	29. Crank-Shaft Bracket.
3. Transfer Piston.	12. Bell-Crank Link.	21. Foot Valve.	30. Crank.
4. Heater.	13. Bell Crank.	22. Discharge Valve.	31. Crank Pin.
5. Telescope.	14. Bell-Crank Bracket.	23. Vacuum Chamber.	32. Heater Bolts.
6. Furnace.	15. Bed Plate.	24. Suction Pipe.	33. Transfer Piston-Rod Crosshead.
7. Gas Burners.	16. Fly-wheel.	25. Stanchion.	
8. Air Chamber.	17. Air Piston Links.	26. Foot-Valve Chamber.	
9. Main Beam.	18. Pump Link.	27. Legs.	

A sectional view of an Ericsson engine equipped with a gas burner (illustration courtesy of Alan Phillips).

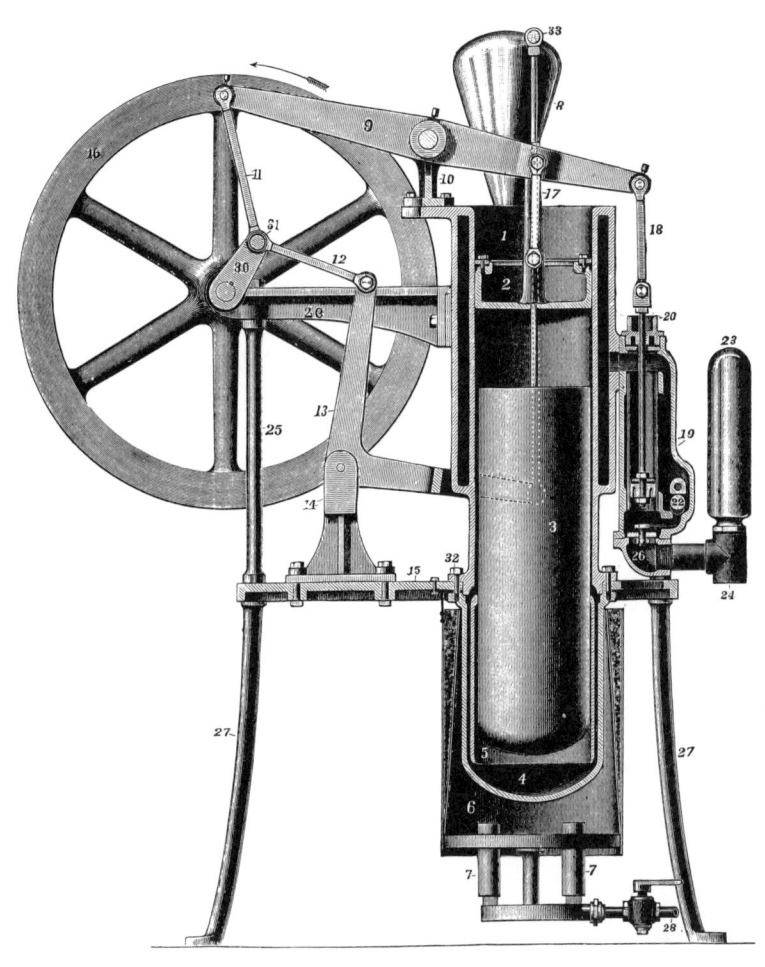

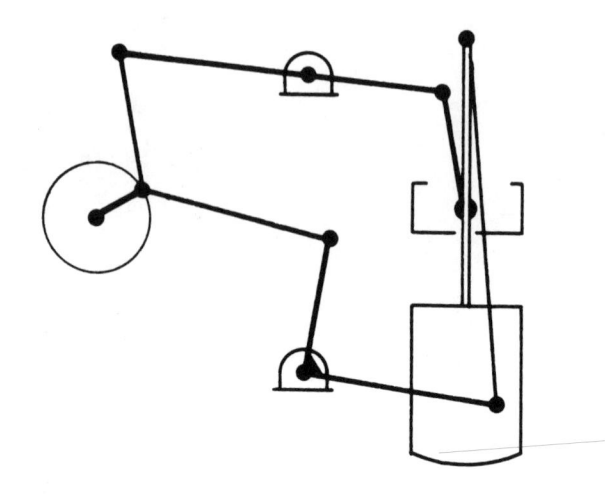

A diagram of the linkage drive system used on the Ericsson engines.

DO YOU USE WATER?

If you have to pump it, why not feel that you are **sure** of a supply? The only **absolutely safe** machine that may be depended upon at all times is a Rider or Ericsson Hot-Air Engine. A **record of twenty years** as an endorsement. If interested send for new catalogue " H."

RIDER ENGINE CO.

86 Lake Street, CHICAGO.   37 Dey Street, New York.

An advertisement from 1915 for Ericsson and Rider hot air pumping engines (illustration courtesy of Bill Perleberg).

This Stirling driven fan available in 1915 featured a single-cylinder type engine (illustration courtesy of Bill Perleberg).

This Fan Runs **Without Electric Current**

Fresh, Cool Air Everywhere

In the kitchen—dining-room—bedroom—on the porch—at the lake—wherever duty or pleasure calls you — you can keep cool and defy the mosquitoes with an

AL-COOL FAN

The mechanism of the fan is simplicity itself, but thoroughly practical and the most economical fan in use. It costs but ½c an hour to run and either denatured alcohol or gas will operate it.

Twelve-inch blades and fan guard both heavily nickel plated. Body finished in black baked enamel. Ornamental as well as useful.

Price Only **$11.50**

Order an Al-Cool Fan today, subject to our 10 days' trial, money-back guarantee.

We want more good dealers. Send for descriptive literature.

AL-COOL FAN COMPANY
10603 Corliss Avenue
CHICAGO, ILL.

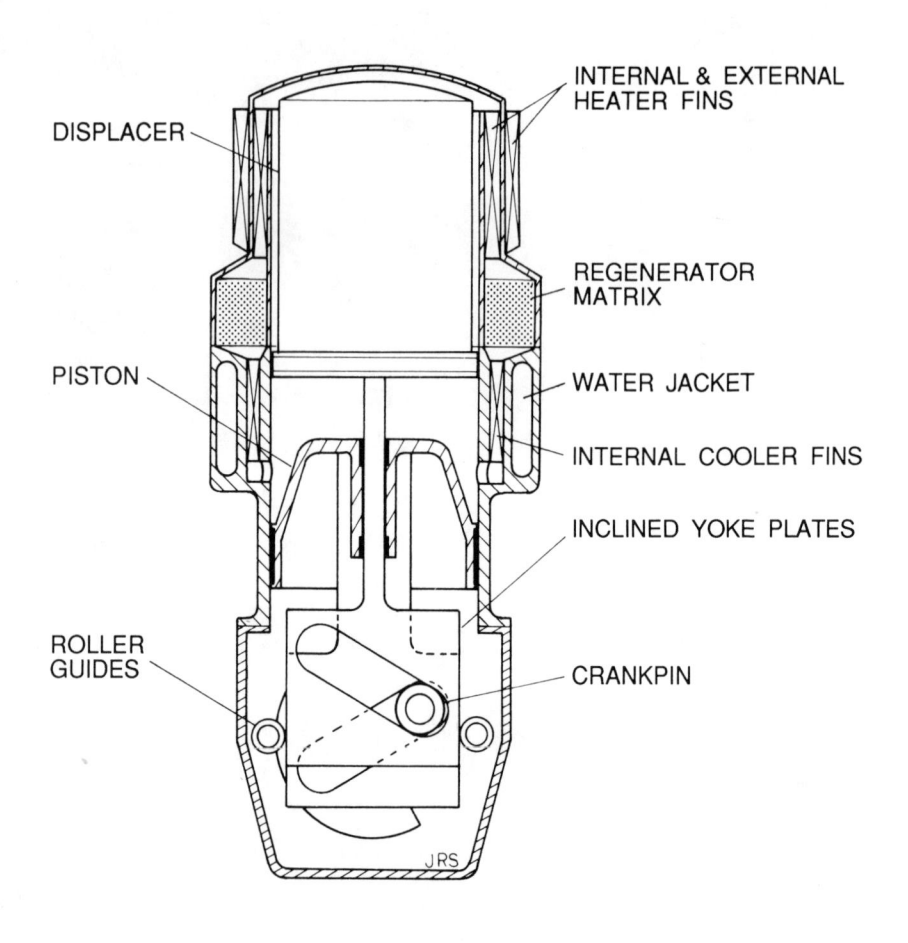

An example of a newer mechanical drive system for single-cylinder type Stirling engines is the "inclined yoke" drive. A scotch yoke with an angled slot connects the piston to the crankpin, and another angled the other way links the displacer to the same crankpin. Each oscillates as the crankshaft rotates, and the angles make a phase difference between the two. This is a simple and compact arrangement.

The Ideal Stirling Cycle

The cycle carried out inside an engine can be represented by recording how the pressure of the working gas changes with its volume. This has been done in Fig. 20 for the ideal Stirling cycle corresponding to figure 19. The horizontal axis represents the total volume of air or whatever other gas is in the engine, and the vertical axis represents its pressure. The trace of the cycle undergone by the working gas is called the *pressure-volume diagram* or "p-V" diagram for the engine.

In the ideal Stirling cycle, there are two constant volume processes where the piston is stationary. These are represented in the p-V diagram by the straight vertical segments. Segment 2-3 is the heating resulting from the displacer transferring air into the hot space in Fig. 19a. Segment 4-1 is the other transfer stroke where the gas is cooled, Fig. 19c. Curve 3-4 is the expansion in the hot space, Fig. 19b. This is carried out with the gas all at the high temperature T_H all through the expansion and so is referred to as an isothermal process. The curve 1-2 is the compression which is also isothermal but at the lower temperature T_C.

A pressure-volume diagram is a concise way to portray an engine cycle. One of its principal values is due to the fact that the area within a p-V diagram represents the work that the engine produces each cycle. This can be proven with the aid of calculus. The heat that flows into and out of the engine working gas can also be indicated on a p-V diagram. This has been done in Fig. 20 for the ideal Stirling cycle. Thermal energy Q_i flows into the engine during the isothermal expansion 3-4. The heat flow out in the isothermal compression is Q_O. Quantities Q_A and Q_B are the heat flows into and out of the engine gas in the transfer strokes. Now if a perfect regenerator is used in the engine, then Q_A and Q_B are continually recycled and so do not come in from the source or escape out to the sink as already explained. This again shows how the regenerator improves efficiency.

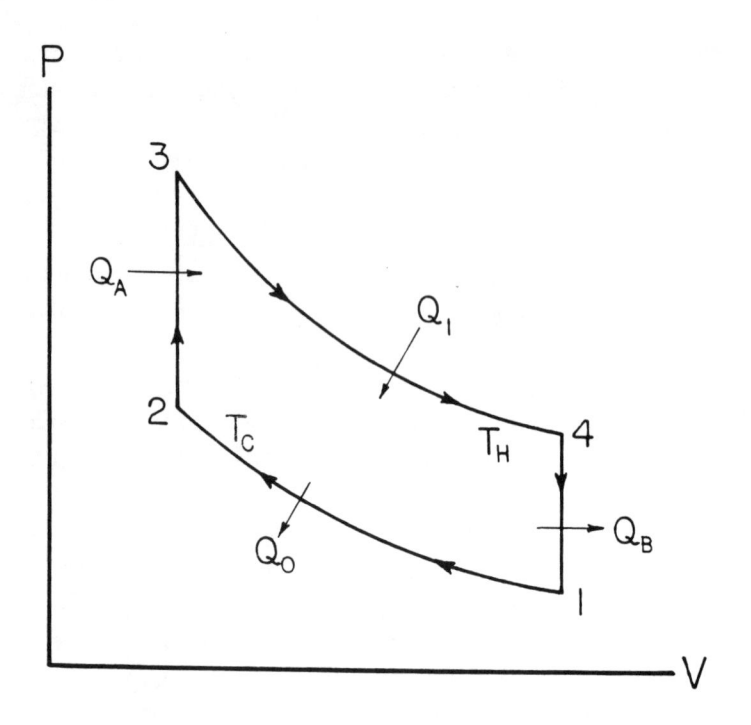

Fig. 20   A p-V diagram for the ideal Stirling cycle. The horizontal axis represents the total volume of the working gas within the engine capsule. The vertical axis represents the pressure of the gas inside the engine. Process 1-2 is an isothermal compression at the temperature T_C. For an ideal gas, the exact mathematical shape of this curve is a hyperbola. Process 2-3 is a constant volume heating of the engine gas during its transfer into the hot section. Process 3-4 is an isothermal expansion at the higher temperature T_H. The final step 4-1 of the cycle is a constant volume cooling of the gas as it is displaced back into the cold section of the engine. The area enclosed by the diagram represents the net work done by the engine in each cycle. Also shown on the diagram are the quantities of heat absorbed or rejected in the four processes.

The work output in a cycle is $Q_i - Q_o$ by the First Law of Thermodynamics. With regeneration, the heat supplied from the source is Q_i, so the efficiency is $(Q_i - Q_o)/Q_i$. A simple calculation using the Ideal Gas Law shows this is equal to $(T_H - T_C)/T_H$ which is exactly the Carnot efficiency described above. By the Second Law, this is the best that any engine operating between these same two temperatures can do. Thus the ideal Stirling cycle with regeneration has the maximum potential thermal efficiency. The Stirling is therefore not only important because of its practical advantages such as its quiet clean characteristics, but is also theoretically important as well. Moreover, it has just recently been proven that the ideal Stirling engine has the highest mechanical efficiency theoretically possible as well. So the Stirling engine has the highest overall efficiency potential of all possible heat engines. Thus the Stirling occupies the key position in the theory of heat engines.

Two-Piston Stirling Engines

In 1875, A. K. Rider of Philadelphia invented an engine which carried out the Stirling cycle in an entirely new way. His system had two pistons instead of one and did not have a displacer. To understand how his engine worked, consider two connected cylinders closed off by pistons as shown in Fig. 21. One cylinder is kept cool and the other is heated at the bottom end. The hot cylinder on the left in the figure is kept cool at its top end because that is where the piston rides. It is difficult to maintain close fits or provide adequate lubrication at high temperatures. The piston extends down into the hot section, but has clearance around it in that region to prevent damage from rubbing hot surfaces.

In Fig. 21a, the transfer stroke is just beginning. The cold piston moves inward and the hot piston moves outward at the same rate. The engine air is pushed through the connecting passage, where the regenerator resides, and into the hot section. At the beginning of this process, most of the air inside the engine was in the cold cylinder. At the end, most of it is in the hot cylinder with higher temperature and

(a) TRANSFER

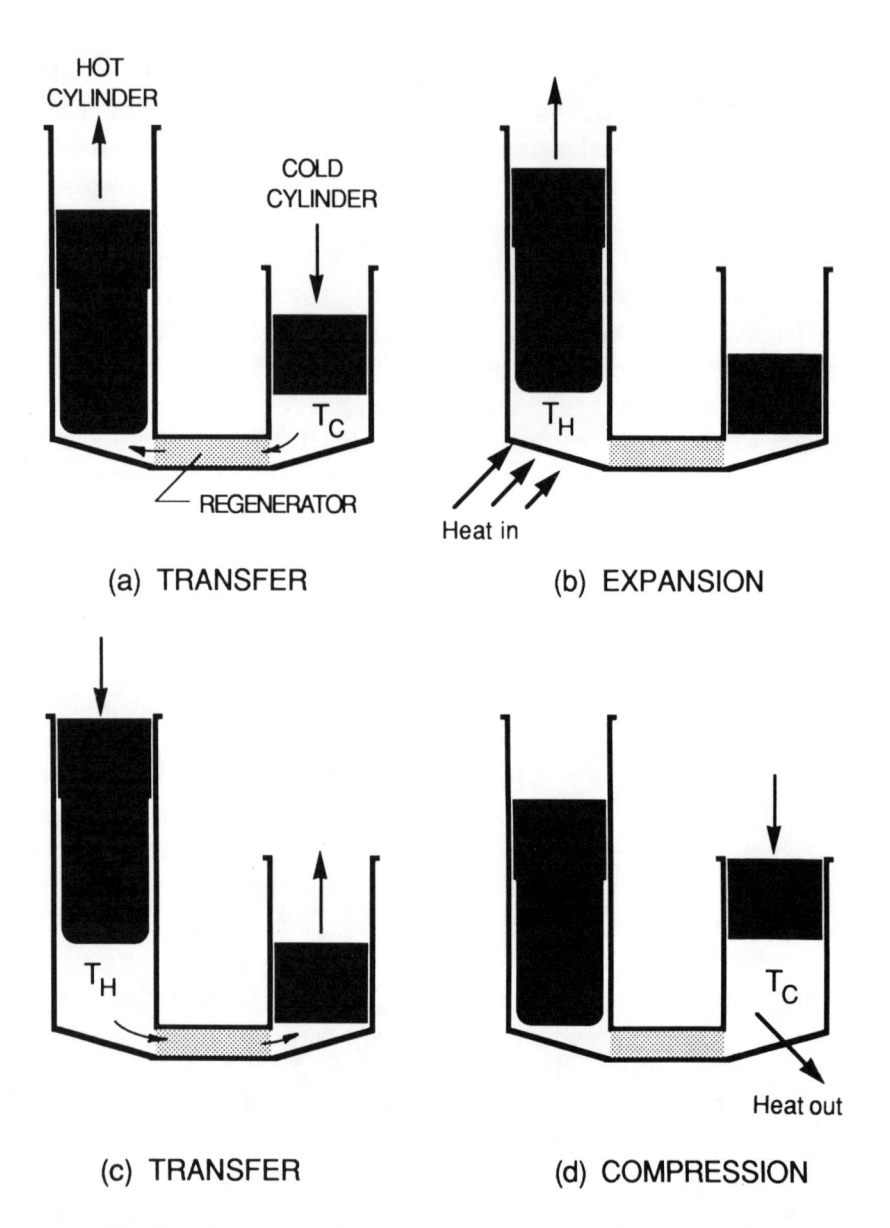

(b) EXPANSION

(c) TRANSFER

(d) COMPRESSION

Fig. 21 Sequence of ideal motions of the pistons of the Rider configuration for a Stirling cycle engine.

correspondingly higher pressure. The expansion stroke follows in Fig. 21b with the hot piston continuing its outward movement. Heat is absorbed from the source in the expansion stroke.

After the hot side piston has completed the expansion stroke, both pistons move together again as in Fig. 21c. The hot one goes in and the cold one moves out. This pushes the hot air through the regenerator into the cold side. Finally the cycle is completed by the inward movement of the cold piston in Fig. 21d to effect the compression. The piston moves about halfway in and heat is rejected in this step to keep the engine air at the lower temperature.

The Rider two-piston or "alpha" arrangement can carry out the ideal Stirling cycle just as well as the single-cylinder piston-displacer arrangement. In practice however, because of the limitations of mechanical drives, the exact ideal motions cannot be achieved. But as in the single-cylinder and split-cylinder cases, ordinary crank drives make good practical approximations. In the original engines, Rider used two cranks on a common shaft with the flywheel in between. They were made and sold in the same era as the Ericsson engines also for water pumping. The Riders had a regenerator located in the connecting passage as in Fig. 21. However, both pistons were made long and the connecting passage was located up higher which enhanced the transfer of heat into and out of the engine. The Riders were very good performers because of these features. Many modern versions of the Rider type Stirling engine have been made on an experimental basis.

KING EDWARD VII.,

the new King of England, uses a Rider Engine in his palace at Sandringham. The Khedive of Egypt has a Rider Engine at Ras-El-Tin palace at Alexandria, Egypt. The Paris Exposition gave the highest Medal of its class to both our Rider and Ericsson Engines.

Rider and Ericsson Hot Air Pumping Engines are appreciated in other countries besides their own apparently.

Catalogue " H " on application to nearest office.

RIDER-ERICSSON ENGINE CO.,

22 Cortlandt Street, New York.   86 Lake Street, Chicago.
239 Franklin Street, Boston.   40 N. 7th Street, Philadelphia.
22a Pitt St., Sydney, N. S. W.   692 Craig Street, Montreal, P.Q.
Teniente-Rey 71, Havana.   Merchant & Alakea Sts., Honolulu.

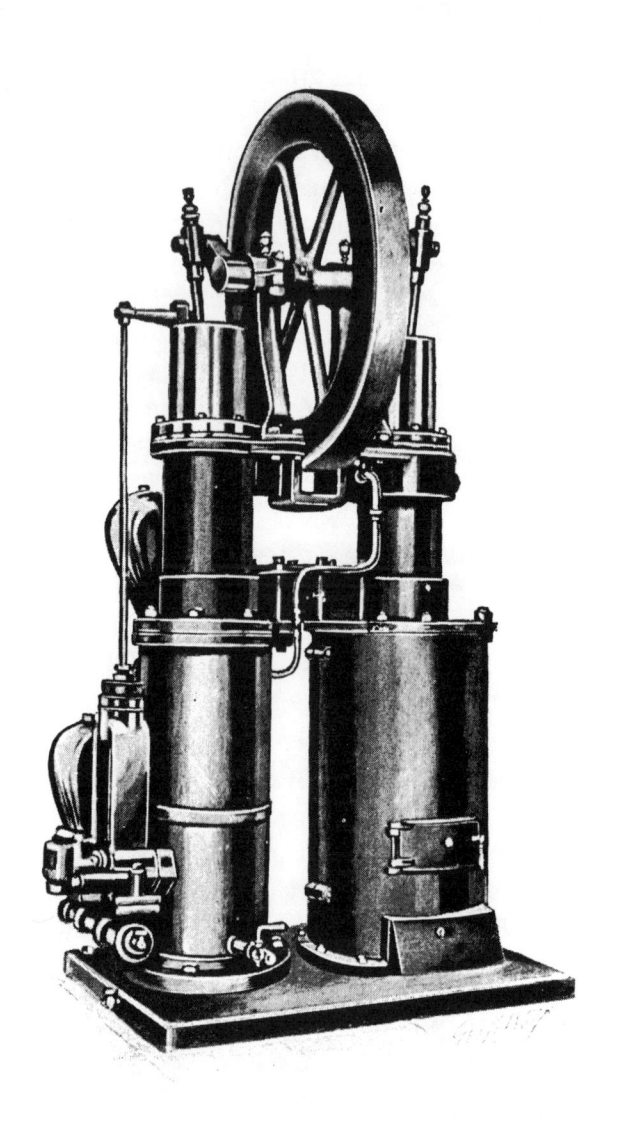

A Rider hot air pumping engine from the turn of the century.

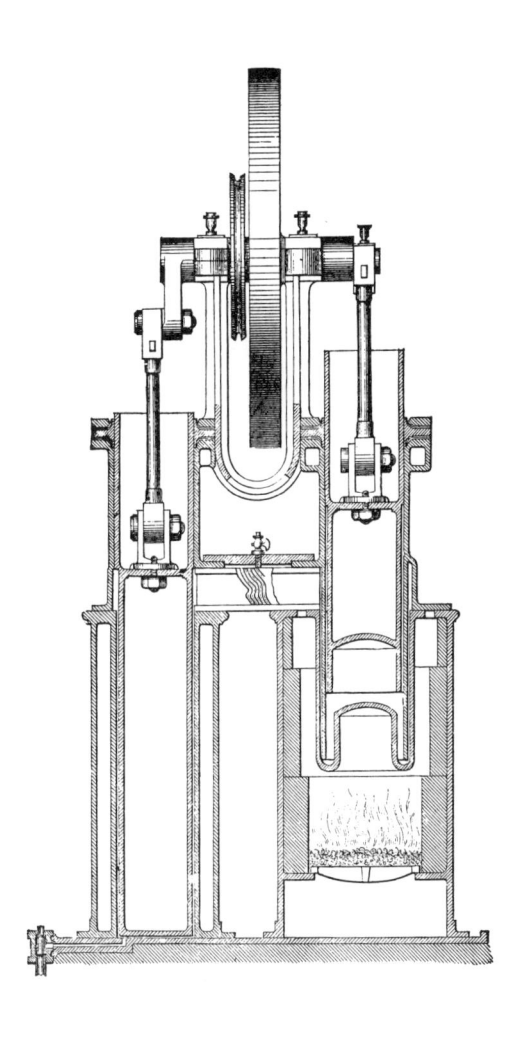

A sectional view of a Rider pumping engine. The long pistons formed a narrow annular space for air flowing into or out of the hot and cold spaces. This enhanced heat transfer and improved performance. The regenerator is located in the connecting passage.

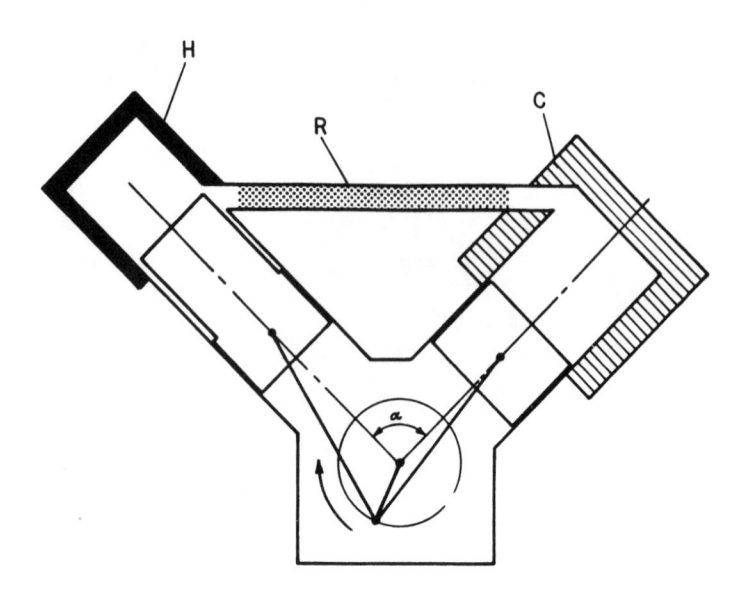

The V-configuration makes a simple two-piston type Stirling engine.

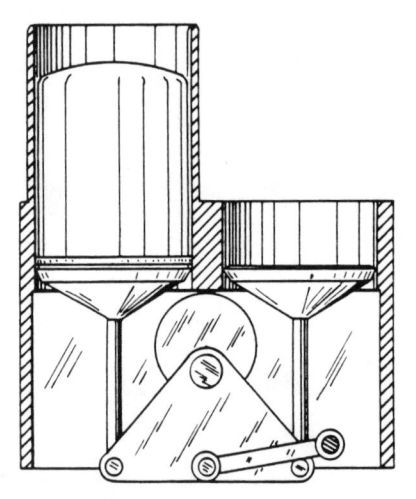

An invention of Andy Ross for the mechanical drive of a two-piston type Stirling engine (illustration courtesy of Andy Ross).

Pressurization

The idea of pressurization was conceived by James Stirling as a means of increasing the power level of his brother's engine. The effect of pressurization is analogous to "supercharging". Pumping air into an engine to raise its average pressure up to say 10 atmospheres means that the engine has ten times more air with which to work. In other words, the engine has the working substance of ten engines inside, and so it behaves more like ten engines than like one. In each cycle, the engine can take in ten times the heat that it did before being pressurized, and so it delivers ten times the work (and also rejects ten times the heat it rejected before to the sink). Therefore, the overall effect is that power can be increased in proportion to the level of pressurization.

In 1843, the Stirling brothers converted a large steam engine into a hot air engine capable of being pressurized. The engine featured twin displacers and a double acting piston 16 inches (41 cm) in diameter with a 4 foot (1.2 m) stroke. This engine was pressurized to about 14 atmospheres average and produced 45 horsepower (34 kW) with an efficiency of 18%. This was an outstanding level of performance for the time. For four years the engine powered all the machinery in the Dundee Foundry Company. However, the engine was plagued by oxidation of the cast iron hot sections of its displacer cylinders. Three times these vessels had to be repaired. Because of the disruption in the work of the company that this caused, the engine was converted back to the more reliable steam operation, even though it was less efficient. Oxidation is not a problem today with stainless steel and other heat resisting alloys available. Current high performance Stirling engine prototypes are highly pressurized, some up to 200 atmospheres. These engines use helium or hydrogen as the working fluid instead of air because these gases have better thermal conductivity and result in lower flow losses. The resulting Stirling engines have power densities well above those of gasoline engines, as well as higher efficiency.

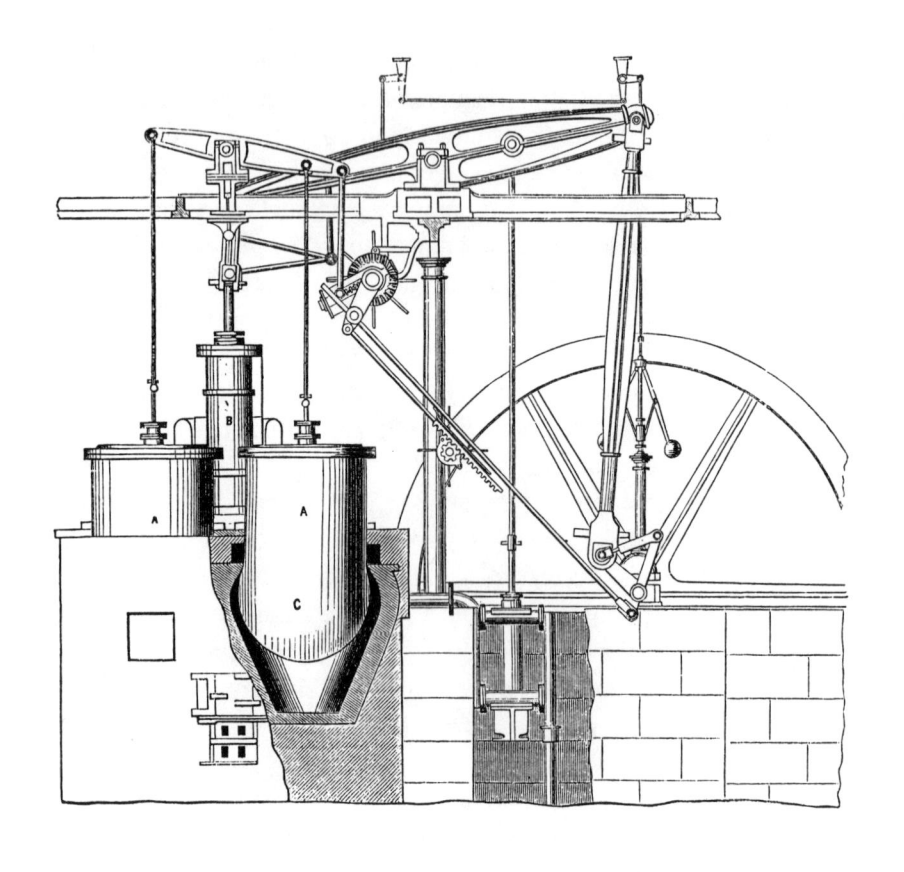

The Stirling brothers' pressurized engine of 1843.

Modern Stirling Engine Development

The modern era of Stirling engine research and development began in 1937 at the Philips Company in Holland. At that time, the company began looking for an engine which could drive a generator and make enough electricity to power the radios that they manufactured. Not much power was needed; a 100 Watts or so would run a lot of radio. But the engine had to be very quiet, long lasting, simple, reliable, and able to use widely available kerosene as a fuel. It was realized by the Philips engineers that the Stirling hot air engine had a lot of potential for this application, especially because heat resisting stainless steels were then available, and a sound scientific understanding of thermal physics and heat transfer had been developed.

Within five years, Philips had developed an amazing little hot air engine, the "Model 10". This single-cylinder engine was about the size of a small lawn mower engine having a bore of 54mm (2.1 in) and a stroke of 28mm (1.1 in). The drive mechanism was based on the Ericsson bellcrank system. Philips engineers cleverly transformed it into a very compact arrangement which permitted the entire drive system to be enclosed in a crankcase and pressurized.

The Model 10 engine produced 2/3 HP (500 W) at a speed of 1500 rpm when pressurized with 5 atmospheres of air. With more pressure, it could be coaxed up to 1 HP at 2000 rpm. With hydrogen instead of air as the working gas, the engine could make 1-1/2 HP. Compared to the large slow moving air engines of the first era, this was phenomenal performance.

Although the development of the transistor later made the need for the initially envisioned small generating outfit obsolete, the early successes inspired interest in developing larger engines for other applications. In 1953, R. Meijer of Philips invented the elegant rhombic drive system for the single-cylinder type Stirling engine. This mechanical drive features twin counter-rotating crankshafts with four connecting rods. It is possible to balance the rhombic drive engine completely and achieve totally smooth operation at high speeds.

A series of rhombic drive engines of increasing size followed, culminating in an in-line four cylinder version having a total displacement of 940cc (58 cu. in). This engine running with helium as the working gas at a pressure of 110 atmospheres produced 90 HP (68 kW), which exceeds the power density of diesel engines. In 1971 the engine was installed in a bus and successfully road tested for over 1500 km. Another engine type pioneered by Philips is the four-cylinder Rider type engine with swashplate drive. Because of its compactness, this was the engine of choice by Ford Motor Company for first testing in an automobile in 1975.

In addition to their innovations in the field of Stirling powerplants, the Philips Co. also developed the Stirling machine into a commercially successful refrigeration machine. The Stirling cycle is a thermodynamically reversible cycle. Each of the four processes of the ideal cycle of Fig. 20 can be performed backwards. If the piston or the displacer is moved the other way, the corresponding thermal process occurs in reverse. So if a Stirling engine is driven backwards, say by an electric motor, the heat flows are opposite to those when it is running as an engine. In other words, the arrows in the basic heat engine diagram of Fig. 10 are reversed. It will absorb heat from a low temperature and deliver it, plus the equivalent of the work received from the motor, at a higher temperature. Thus it acts as a refrigerator or heat pump. In fact Stirlings are very effective heat pumps for operation in the cryogenic range, below 100 Kelvin. Very early on Philips began marketing Stirling cryocoolers for liquefying air and various industrial gases. These machines just hum away and condensed liquid air drips off the cold end. Medium size Stirling coolers are now being developed by other companies for chilling semiconductor circuits; at a temperature of about 80 K (the temperature of liquid nitrogen at atmospheric pressure) computer chips work twice as fast and last up to four times longer than under normal conditions.

Almost immediately, the achievements of Philips interested others in the many possibilities of the new breed of Stirling machine, and research and development projects of all kinds sprouted up and have continued up through the present. By now there is a long list of companies, government agencies, and universities all over the world that have worked on various aspects of Stirling engine research and development. Experimental Stirling engines have been designed and tested as clean and quiet powerplants for automobiles, trucks, buses, boats, and submarines. Small Stirling engines have been tested for powering artificial hearts. Miniature Stirling cyrocoolers are in regular use chilling sensors in night vision equipment. Stirling coolers are being investigated for freon-free refrigerators. Stirling engine driven Stirling heat pumps are being developed for home heating; these units could deliver up to 30% more heat to a home than the fuel burned would give alone, which would make a huge impact on conservation efforts. Stirling engines are now being developed for solar operation to drive generators and make electricity for space stations, planetary bases, and for our everyday use on earth as well. The quiet call of an engine operating in greater harmony with nature and the human condition seems irresistible, and perhaps the Stirling engine will soon realize some of its promising potential.

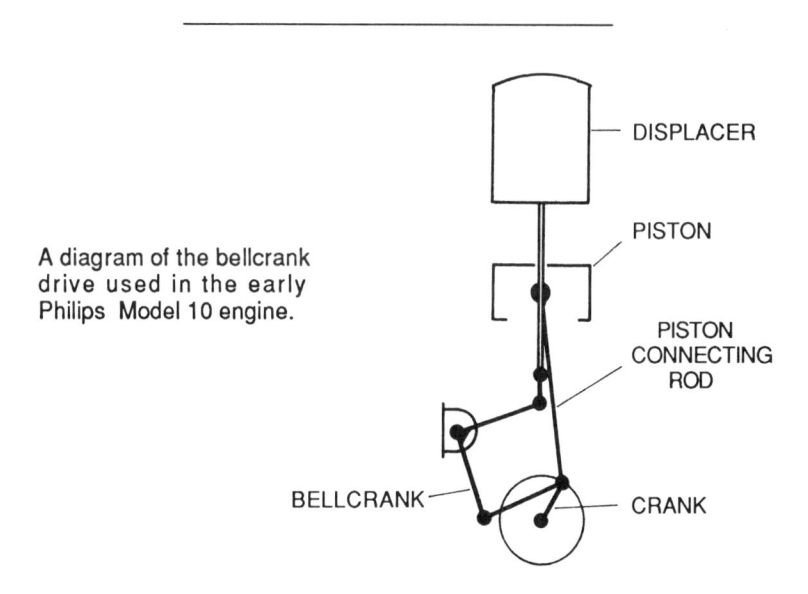

A diagram of the bellcrank drive used in the early Philips Model 10 engine.

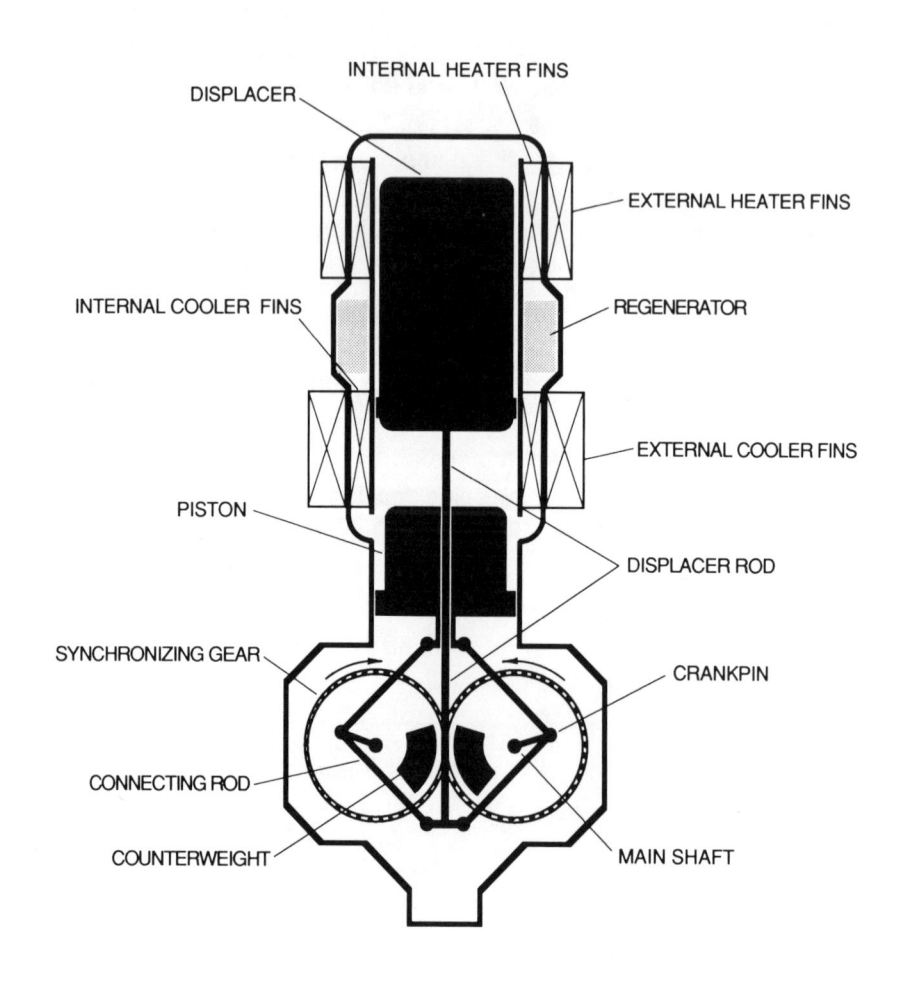

The rhombic drive mechanism is shown here in an engine designed with finned heat transfer surfaces inside and outside the engine cylinder.

In 1979, Sunpower Inc. designed and built a prototype rhombic drive engine shown here pumping water. It featured finned heat transfer surfaces similar to those shown in the diagram above. The engine bore is 8" (20cm) and the stroke is 3.1" (8cm). Running with air at atmospheric pressure, the engine produced 1.2 Hp at 1100 rpm.

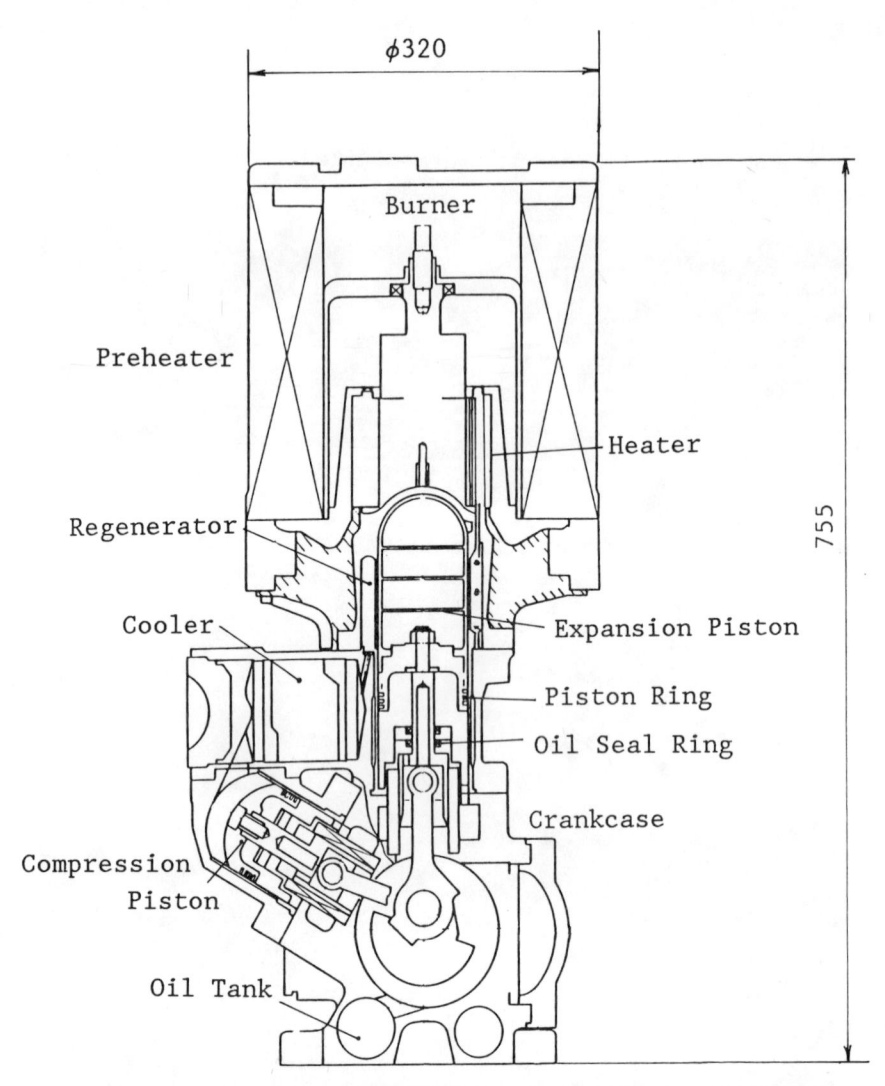

An experimental two-piston type Stirling engine designed and built by the research labs of Toshiba in 1987 for driving a residential heat pump. The V angle is 60°, but two crankpins are used (only one is shown) set 40° apart to give a total phase angle of 100° which is optimum for a high temperature two-piston Stirling engine. The engine is charged with helium at 60 atmospheres pressure, and the engine produces 3 kW at 1500 rpm burning natural gas as fuel (illustration courtesy of Dr. N. Kagawa).

A 1979 AMC Spirit equipped with an experimental Stirling engine powerplant, the "P-40". The car was capable of burning gasoline, diesel, or gasohol. The higher efficiency of the Stirling engine promised less pollution, 30% better mileage, and the same level of performance as the car's standard internal combustion engine. The 4-cylinder Rider-type Stirling engine used hydrogen as the working gas at a mean pressure of 2200 psi (150 atm) with a total piston displacement of 492cc (30in^3). The engine produced 72 Hp (54kW) at 4000 rpm.

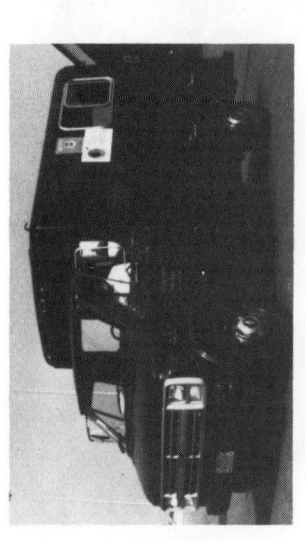

PHASE I—AIR FORCE C-30 VAN

PHASE II—AIR FORCE D-150 TRUCK

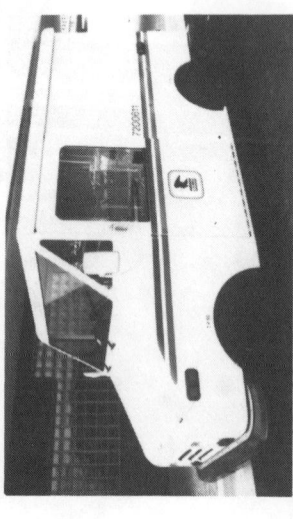

PHASE III—UNITED STATES POSTAL VAN

As an adjunct to their research in developing the Stirling engine for use in space, NASA has extensively tested an upgraded automotive Stirling engine in the three vehicles shown here. Fuels tested include unleaded gasoline, JP-4 aircraft fuel, and DF-2 diesel fuel.

A view of the Stirling powerplant used in the above vehicles. The large round object is the outside of the combustion chamber which heats the ends of the four cylinders of the engine. Because the combustion is continuous, the pollution levels of a Stirling engine are dramatically low.

The intricate array of heater tubes inside the combustion chamber of the automotive Stirling engine present a large surface area to rapidly pick up thermal energy from the combustion process. In the upper area these tubes carry fine fins to further enhance heat transfer.

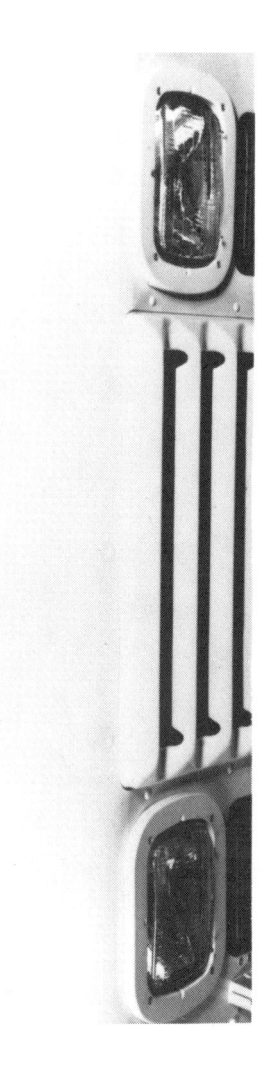

A view of the Stirling engine powerplant under the hood of the postal van.

This 36' (11 m) diameter dish of mirrors is parabolic in shape, and sunshine is reflected from its surface and converges to a single spot, the focal point, 20' (6 m) above the center of the dish. Positioned there is the hot end of a Stirling engine (United Stirling model 4-95) which is coupled to an electric generator. The engine is cooled by a water-radiator system. The entire unit is mounted on a pivoted pedestal so that the sun can be tracked throughout the day. The unit produces 25 kW of electricity at a record solar-to-electric efficiency of 30%.

Bibliography

The following books are a good next step for learning more about what has been introduced in this book. Each will provide more information on various aspects of Stirling engines ranging from their history to their mathematical analysis. These books provide many more references to books, articles, and papers on Stirling engines for further study.

Fenn, John B.　Engines, Energy, and Entropy , Freeman, San Francisco, 1982.　A delightful introduction to classical thermodynamics. Easy to read with clear explanations of mathematical methods used.

Hargreaves, C. M.　The Philips Stirling Engine , Elsevier, Amsterdam, 1991.　A comprehensive technical story of the research at Philips which gave birth to the modern Stirling.

Kolin, Ivo.　The Evolution of the Heat Engine , Longman, London, 1972.　A superbly illustrated technical history of the heat engine with substantial coverage of the Stirling.　Comprehensive comparison of major engine types in the context of thermodynamics.　Much is accessible to beginners. Reprinted by Moriya Press, 1998.

Mott-Smith, Morton　The Concept of Energy Simply Explained , Dover, New York, 1964.　A non-mathematical description of the principles, applications, and historical development of thermodynamics.

Reader , G. T. & Hooper, C.　Stirling Engines , Spon, London, 1983. A complete and solid coverage of the fundamentals of Stirling engine technology especially aimed at engineering students.

Ross, Andy　Stirling Cycle Engines , Solar Engines, Phoenix, 1977. A well-illustrated non-technical story of the Stirling engine .

Senft, J. R.   <u>Ringbom Stirling Engines</u> , Oxford Univ. Press, New York, 1993. A technical treatment of the Ringbom type of Stirling engine.

Van Ness, H. C.   <u>Understanding Thermodynamics</u> , Dover, New York, 1969. A technical but mathematically simplified explanation of the laws of thermodynamics and their bearing on heat engines.

Walker, G.   <u>Stirling Engines</u> , Clarendon Press, Oxford, 1980.
The standard technical work in the field. Complete coverage.

Walker, G. & Senft, J. R.   <u>Free Piston Stirling Engines</u> , Springer, Berlin, 1985. A technical treatment of free-piston and related types of Stirling engine.

West, Colin D.   <u>Principles and Applications of Stirling Engines</u> , Van Nostrand, New York, 1986. An introduction to the Stirling for those with an engineering background. Especially good treatment of basic methods of analysis.

Sources

New Machine Company; 12121 NE 66th Street; Kirkland WA 98033. Model Stirling engines and books.

Bailey Craftsman Supply; P. O. Box 276; Fulton MO 65251-0276. Stirling engine books, kits, tools, and supplies.

Stirling Machine World; 8880 N. Duskfire Drive, Tucson AZ 85737-2366. Quarterly newsletter, books, reprints of papers, model engines, and videos.

Small Parts Inc., 13980 N. W. 58th Court, Miami Lakes FL 33014. Suppliers of materials, mechanical components, tools, and books.

ALSO FROM MORIYA PRESS

An Introduction to
Low Temperature Differential Stirling Engines

James R. Senft

The story of the new breed of Stirling engines that can run on remarkably small temperature differences.

Describes how to make a model engine that will run just on the heat of your hand.

An Introduction to Low Temperature Differential Stirling Engines covers the latest new development in the field of Stirling engines. This 88-page book introduces the fascinating world of Stirling engines that can operate on hot water, ice, sunshine, or even on the warmth of your hand. Developed over a decade of university research, low temperature differential Stirlings have been refined to operate wherever there is a temperature differential of just a few degrees. Written by one of the original researchers in the field, the subject is explained from its beginnings in a clear, logical, and complete way, and the design and operating characteristics of these engines are related to the basic facts of heat engine physics. For readers interested in making one of these remarkable engines, the book includes complete dimensioned drawings and directions for making an engine that will run effortlessly just resting on your hand for heat. The book is entirely self contained and accessible to the raw beginner, but the expert will find much new material in this volume including a long list of references for further study. The author is a university professor with over two decades of research experience in heat engines. He has written and lectured extensively on Stirling engines, and served as consultant to universities, laboratories, and industrial organizations on engine development.
88 pages.   ISBN 0-9652455-1-9.   **$13.95** postpaid.

Order from your favorite bookseller or direct from
Moriya Press   P. O. Box 384   River Falls WI   54022

ANOTHER BOOK FROM MORIYA PRESS

The Evolution of the Heat Engine
Ivo Kolin

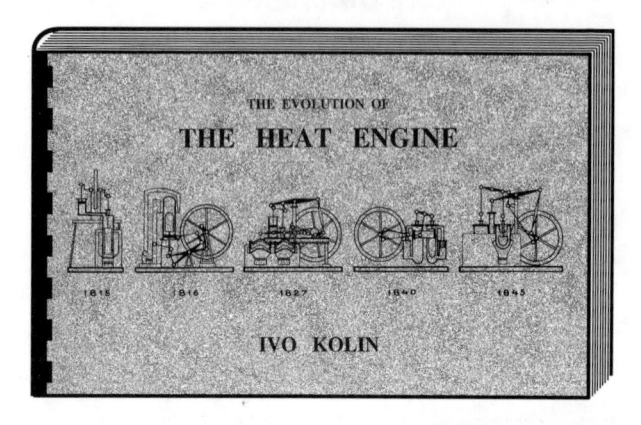

The Evolution of the Heat Engine traces the development of the thermal machines of the present from their origins in the past. The story is told in this volume through superb drawings and diagrams. The illustrations are all the original work of the author prepared especially for this book. All are accurately based upon original historical sources and are drawn to show all the essential working details in an exceptionally clear way. The book also presents technical data on each engine, including overall dimensions and many test results. Throughout the book, the thermodynamic principles that underlie all these devices are explained and illustrated by pressure-volume diagrams, charts, and thermodynamic calculations. There is also a chapter on thermodynamic theory followed by an insightful summary chapter which compares and classifies heat engines according to heat transfer, open or closed cycles, working substance, efficiency, and specific power. The book is an invaluable resource for scientists, engineers, and historians of technology. Anyone curious about the how and why of engines, heat pumps, or rockets will find answers in this book. For model makers there is a wealth of ideas and inspiration in these pages for new projects. Those interested in tapping alternate energy sources will find this book an excellent background guide. The author, Dr. Ivo Kolin, is Professor Emeritus of Thermodynamics at the University of Zagreb. He is the inventor of the low temperature differential version of the Stirling engine and enjoys world-wide recognition as an expert in the field of heat engines. Prof. Kolin is also a gifted illustrator which makes this book a uniquely masterful presentation of the story of heat engines.
   Large format.   106 pages.   ISBN 0-9652455-2-7.   **$21.95** postpaid.

Order from your favorite bookseller or direct from
Moriya Press   P. O. Box 384   River Falls WI   54022